CONTENTS

KU-075-440

Peter J. Williams
Carleton University
Ottawa

PIPELINES AND PERMAFROST:

physical geography and development in the circumpolar North

Longman
London
and New York

Longman Group Limited London

*Associated companies, branches and representatives
throughout the world*

*Published in the United States of America
by Longman Inc., New York*

First published 1979

British Library Cataloguing in Publication Data

Williams, Peter
   Pipelines and permafrost. – (Topics in
   applied geography).
   1. Gas, Natural – Arctic regions – Pipe lines
   2. Petroleum – Arctic regions – Pipe lines
   3. Frozen ground – Arctic regions
   I. Title II. Series
   621.8'672   TN880.5   79-40298

   ISBN 0-582-48576-2

Printed in Great Britain by
McCorquodale (Newton) Ltd., Newton-le-Willows, Lancashire.

TOPICS IN

1982

PIPELINES AND
PERMAFROST

TOPICS IN APPLIED GEOGRAPHY
edited by Donald Davidson and John Dawson

LIST OF PLATES

This book is not only about pipelines and permafrost. It is as much an account of the erratic development of our knowledge of a part of the earth's surface, of the demands for that knowledge as great technological projects are expounded, and of the buffeting these projects receive as the deficiences of understanding become apparent.

When the major northern oil and gas pipeline proposals reached the design stage in the early 1970s, there were surprising gaps in our knowledge of the behaviour of freezing soils, the material upon which the pipelines would be built. The twists and turns in the development of designs as problems came to light during the next few years, make a fascinating, sometimes disturbing, story. It is one of interest not merely to the professional scientist or engineer. The technological difficulties associated with cold regions are real enough, and the possible consequences of failure so great, that these matters should be of concern to others than the pipeline companies themselves.

During the last two decades interest in scientific and technological achievements has waned, giving way to concern about how man utilises his obvious abilities in these respects. The building of a pipeline does not carry the same aura of excitement which earlier generations found in such construction feats as the Roman aqueducts, the great wall of China, or even the transcontinental railways. Instead, public interest has centred on the economic desirability, the social and political consequences, and the disturbance of the natural environment which major technological undertakings represent.

The apparent simplicity of a big tube in the ground to carry gas or oil thousands of miles is deceptive, however, and particularly so when that ground experiences the extreme cold associated with Siberia, northern Canada and Alaska. The challenge that the natural conditions of the earth's surface present to the building of northern pipelines must be understood, before the broader implications of the pipeline projects can be properly considered. The science from which designs are developed, should be a basic element in the democratic examinations and inquiries, such as those to which the northern pipeline proposals have been subject, and which society increasingly demands of major engineering projects. Indeed, scientific questions have been part of such public enquiries although insufficiently so. The social and environmental issues frequently arise because of, or are modified by, the technological and scientific constraints. But, for the most part, the geotechnical aspects of northern pipelines have gone unheeded or unnoticed by the media, alongside the more easily described topics of native rights, nature conservation and similar aspects.

This book sets out to explain the nature of the special problems the cold northern environments pose for large pipelines. The language used is, for the most part, not that of the specialist scientist or engineer. There are no equations in the body of the text and technical or scientific terms are used sparsely and with careful explanation. A few of my scientific colleagues, who are specialists, will doubtless feel my approach is superficial, but the book is not written for them. There are many earth scientists who will find the topic interesting, even if presented in quite simple terms.

I have not hesitated to comment on what some might regard as the mere sociology of the scientific research and development processes. To assume that this aspect is not of relevance to the central issues is misguided. The occasional inclusion of comments which could be construed as 'political' has, I believe, in no way influenced an accurate and balanced view of the scientific questions. In writing for those whose interest is in the economic, social or environmental implications of the pipeline projects, it is important to represent the scientific background accurately and clearly. In this respect the ephemeral impressions recorded by the journalist may be as limited in their usefulness as is the expertise submerged in the private language of technical reports. The task has been eased by the essentially elementary nature of the scientific principles involved (and so long overlooked), and by the interest of the story of key research findings, and their relationship to particular people, places, and events.

If politics is too important to be left to politicians, surely science and technology are too important to be left to the scientists and technologists. Least of all should the interpretation of scientific and technological problems be left to those who have a vested commercial interest in projects of the magnitude of the northern pipelines. This small book, it is hoped, will stimulate broader interest and a more balanced view of a largely overlooked aspect of these giant and consequential undertakings.

My involvement with the geotechnical problems of pipelines in permafrost came about as a result of my experiences as a scientist and teacher. In turn, I must warmly acknowledge my general indebtedness in the writing of this book, to my colleagues and students in the Geotechnical Science Unit, Geography Department, Carleton University. There are many other colleagues in the scientific and geotechnical fields who have also provided information, although neither I nor they may have realised, at the time, that it would be presented here. While omitting their names, I hope that my appreciation of our friendly and collegial conversations will be understood.

I am grateful for advice specifically relating to the text, to Professor P. E. Uren, Carleton University; to Mr. Roland Wahlgren and Mr. Max Perchanok, respectively past and present graduate students in physical geography at Carleton; and to my father J. G. Williams – whose interest has had much to do with making the text readable also for the non-specialist.

Peter J. Williams
Ottawa
Canada

March 1979

ACKNOWLEDGEMENTS

We are grateful to the following for permission to reproduce copyright material:

Academic Press Inc., T. Czudek and J. Demek for Fig. 2.2 modified from Fig. 9 p. 111 Czudek and Demek 1970 – *Quaternary Research 1*; Alyeska Pipeline Service Company for Figs. 4.3, 4.4, 4.5 and 4.6 based on Figs. from p. 7 – *Summary Project Description of the Trans-Alaska Pipeline System*; American Meteorological Society for Fig. 2.3 modified from Fig. 4 Weller and Holmgren 1974 – *Journal of Applied Meteorology B*; Department of Energy Mines and Resources, Canada for Fig. 6.5 part of map 115 St. Elias – *National Topographic System*. Reproduced with permission; National Research Council of Canada and R. J. E. Brown for Fig. 4.2 modified from Fig. by R. J. E. Brown, 1968 – *Permafrost Investigations in Northern Ontario and North Eastern Manitoba Tech. Paper 291*; University of Toronto Press for Fig. 1.2 modified from Fig. 6 p. 11 and Fig. 1.3 modified from Fig. 4 p. 8 R. J. E. Brown 1970 – *Permafrost in Canada*; John Wiley & Sons Inc. for Fig. 5.4 from Fig. 6 p. 354 Burt and Williams 1976 – *Earth Surface Processes 1*.

CHAPTER 1

THE CHALLENGE

Most homes in North America are heated by gas or fuel oil, which has been transported great distances by pipeline. 'North Sea gas' is a household expression in the United Kingdom. Modern industrialised nations depend in large measure on these fuels for energy, and stability of supply is a major factor in international politics. Natural gas alone provides about a third of the total energy requirements in America,[1] and the U.S. Federal Energy Regulatory Commission has commented[1] 'when shortages of natural gas occur, as during the winter of 1976–77, the effects are profound hardship and danger for individuals and substantial economic disruption for the economy'. Yet the linkages for this mainspring of modern society, the pipelines connecting wellhead and consumer, are usually taken as much for granted as the pipes which carry water to our homes. This attitude is unjustified because development has extended pipeline operations to remote and inhospitable places, while the pipes themselves have become large in diameter. The associated technological demands, the scale of construction and the often huge costs, bring with them a host of dangers, commensurate only with the increasing dependence of society on oil and gas. The history of oil and gas pipelines goes back further than the recent decades during which the pipeline industry has come to carry such potential and responsibility in respect to human well-being.

1.1 OIL AND GAS PIPELINES: EARLY DEVELOPMENT

Oil has been transported by pipeline for more than a hundred years. The first such pipeline dates to 1861, was apparently of wood, and about 10 cm in diameter – it carried oil some 10 km in Pennsylvania.[2] By 1874 a pipeline 100 km long carried oil to Pittsburg. By 1878 a line over the Allegheny mountains had been proposed. In spite of the doubts expressed, the line proved successful, and thus for the first time the special difficulties of a particular terrain had been overcome. The history of gas pipelines is even older; for example Genoa had gas street lighting in 1802. The pipes necessary for such a system were rudimentary. The network of natural gas or coal gas distribution pipes of many big cities by the end of the last century, involved only small diameter pipes, usually of iron. But in the early part of this century, pipe sizes increased from 3 to 6 inches and even to 19 inches (30.5 cm) diameter.[3]

The difficulties of transporting gas and oil by pipeline increase with the size of the pipe, and so do the adverse consequences of a rupture, or leakage. During the Second World War relatively large diameter pipelines were introduced, with the 'big-inch' (diameter 24 inches) running 2 000 km, from Texas to Pennsylvania. In itself, a pipe of that diameter to carry fluid was not a very novel or remarkable achievement. The Romans constructed much larger aqueducts, of which one of the most famous is that crossing the Gard river near Nimes in France, supported on a magnificent arched structure – the Pont du Gard. Today tourists walk through the conduit high above the river.

⌈The explosive nature of gas or the polluting nature of spilled oil, however, and the necessity of traversing particularly inhospitable parts of the earth's surface are twin challenges. Like the world's present pipeline networks themselves, these challenges are extensive but not immediately obvious. From some 11 000 km in 1900, 40 000 by 1920 and perhaps 100 000 by 1940, the natural gas distribution system in the United States involved about 280 000 km by the mid-1970s.⌊For the most part, the pipelines are buried and not visible at the ground surface⌉ Diameters of 14 inches to 30 inches are common. Pipelines constructed under particularly difficult or novel conditions have generally been of smaller diameter than elsewhere. The pipeline 'under the ocean' (PLUTO) constructed during the Second World War to carry oil from Britain to the Continent was a mere 3 inches.

It is only since around 1960 that the large diameter pipes, 36 inches, 42 inches and even 48 inches, have come to dominate in the transport of oil and gas from highly productive but remote wells. One of the remarkable technological achievements of this century must surely be the tapping of the resources below the North Sea by this means. A passing knowledge, gained from television, newspaper or film, of the massive rigs and platforms battered by North Sea storms, of the steel pipes which slide like giant spaghetti from pipe-laying boats, and of the successful installation of well pipe hundreds of metres below the sea bed, is sufficient to make us wonder whether any part of the earth's surface could be more challenging.

1.2 PIPELINES FOR COLD REGIONS

Major pipelines have been successfully constructed in very hot regions, in North Africa, the Middle East and elsewhere. The completion of the 30-inch Trans-Arabian pipeline in 1950 was a notable achievement. A highly significant part of the North American reserves of oil and gas lie by contrast, in the cold, farthest north part of the continent, particularly along the northern coast of Alaska, in the Canadian Arctic Islands and to some extent in the north-west of mainland Canada. The existence of this oil and gas has, for the most part, been established only in the last 10 to 15 years. These regions have much permafrost, that is, ground which stays frozen throughout the year, and they are characterised by terrain conditions of great diversity, different from those of more temperate regions. Thus, the latest challenge for the world's petroleum industries is found in northern North America and in Siberia. In these regions pipeline construction and operation has only just begun, in the sense that the extent, impact and importance will be substantially greater in the last quarter of the twentieth century. During the Second World War, the Canol pipeline, some 1 000 km of 4-inch and 6-inch pipe, carried oil from Norman Wells to Whitehorse in Canada, and on to Fairbanks (Fig. 1.1). It was laid on the ground surface, and crossed the mountains of the Yukon. Some 25 000 people were involved in its construction, and the cost was $134 million. But 'very little oil reached Whitehorse and when the war ended the pipeline was dismantled'.[4] Only one major oil pipeline has been completed in the permafrost region of North America – the Trans-Alaska 'Alyeska' line, and this is considered in detail in Chapter 4.

The traveller arriving in certain northern parts may be struck by the barrenness of the landscape, although if he is in one of the larger settlements such as Fairbanks,

Alaska; Yakutsk in Siberia; or Inuvik, at the delta of the Mackenzie in north-west Canada, he may, depending on his own cultural background, find the urban surroundings familiar. He may be surprised to find widespread forests, even at quite high latitudes. All in all, there is often little to indicate immediately that these cold regions pose any great or unique problem to the pipeline engineer or, indeed, to civil engineering generally.

The skill and expertise required in modern gas or oil pipeline technology is not widely realised – most imagine the construction procedure (if they think of it at all) as being akin to laying sewers. In fact, the highly complex techniques must be constantly adapted to the nature of the pipeline's surroundings. More influential and informed opinion, however, has also tended to regard the technological feasibility of major pipelines in the northern land regions as a foregone conclusion. Major pipeline companies, who should know best, have sought and received government approval to proceed with a major gas pipeline – the Alcan line (running from fields on the north coast of Alaska, through the southern Yukon, and on to California), albeit at this stage with important qualifications and conditions. There are, as yet, no operational Canadian or American large-diameter gas pipelines in the permafrost regions.

There are fundamental differences between gas and oil lines, which affect the relationship of the pipe to its surroundings. Thus substantially different designs and construction procedures are necessary. Whether the currently-proposed Alcan gas pipeline, down through Alaska and the Yukon, will ultimately be regarded as a success or a partial or even a total failure, may depend on the criteria for success. In one sense, the success of the Alyeska oil line to date is a qualified one. Its cost, about $7 billion,* exceeded that predicted by $6 billion,[5] an increase hardly ascribable to inflation alone. It has only been in operation for a small fraction of its proposed lifespan, and the total cost of maintenance remains to be seen. The effects of this pipeline on the economy and administration of the State of Alaska as well as directly on the lives of many Alaskans, are profound.

In this book we are not primarily concerned with the economic arguments for or against these colossal projects, although it may be noted that decisions by the President of the United States and of the Government of Canada indicate that they accept the desirability of gas transport by the Alcan pipeline to alleviate expected energy shortages towards the late 1980s. The topics to be discussed bear only indirectly on the questions of native rights. Conservation and the question of how much disturbance of the environment is acceptable, are also secondary considerations in this book. Instead, the importance of the physical environment with respect to the hazards and difficulties it presents for major northern pipelines will be analysed. Of course, the economic and social effects, and environmental disturbance arising from the pipelines are closely related matters. The conditions for the success of pipeline projects will be considered in the light of this analysis. Firstly, however, we should briefly consider what we mean by 'success'.

1.3 PIPELINES AND THE PUBLIC INTEREST

For the present purpose, 'success' for a gas or oil line can be defined as the regular and continuous supply of gas or oil as proposed, with construction and maintenance costs approximating those predicted, and with the incidental effects on its surroundings

---

*1 billion = one thousand million = 10^9. In British usage until recently the billion was always one million million = 10^{12}.

(environmental degradation), being little greater than envisaged at the final planning stages.

The record of giant engineering undertakings of the past, if scrutinised in this way, is not very good. We have only to look at the cost overruns[6] of Concorde, or at the fairly dismal record of many pioneering developments in the last century, the transcontinental railways, for example. The redeeming feature of many was, of course, their pioneering and necessarily experimental nature. At this stage of man's technological and scientific development, the value of an unsuccessful gas pipeline as an 'experiment' appears small.

Alternatively, there could be some higher costs than predicted, some more difficulties and adverse effects, such that there might be dispute subsequently, as to whether the construction of a particular gas (or oil) pipeline was necessary, timely and correct. There must, however, be some hypothetical point at which most thinking people would agree that 'this was a failure'.

Canadian and American proposals for pipelines in the North, together with the importance of the whole oil- and gas-supply question has led to considerable public debate, a number of public or semi-public inquiries and a lot of newspaper and television coverage. In Canada, a full-scale inquiry under Mr Justice Berger, into proposals for a gas pipeline down the valley of the Mackenzie represented a milestone in democratic involvement in what might have been the ultimate in entrepreneurial activity.[7] The Berger Inquiry placed great emphasis on evidence concerning the social and environmental impact of a proposed gas pipeline – and concluded that it should be delayed for 10 years. In common with the regulatory bodies, the Inquiry was also presented with a mass of evidence relating to the geotechnical difficulties and their solution. Justice Berger pointed out in his report that serious geotechnical problems arising from the cold environment and frozen soils, remained unsolved.

The Federal Energy Regulatory Commission (F.E.R.C.)[8] in the United States and the National Energy Board in Canada are regulatory bodies whose functions include authorising construction of gas pipelines. The National Energy Board has said that the issuing of a certificate for the eventual construction of the Alcan line should be conditional upon problems relating to frost effects being solved.[9] But the Canadian Federal Parliament, in September 1977, approved construction of the pipeline in principle, requiring only that the newly-established Canadian Northern Pipeline Agency[10] monitor the particular design solutions and construction operations. Parliament, in fact, was acting in response to the recommendation of the American body, the F.E.R.C., that a pipeline through Canada was desirable if the appropriate agreement could be reached with the Canadian Government.[11] The F.E.R.C. recommendation also made reference to technical problems relating to frozen ground which it concluded would be solved. There has been little public discussion of what the technical problems are.[12] Issues involving the rights and well-being of native peoples are justifiably more newsworthy. The efforts of conservationists with respect to landscapes, flora and fauna most North Americans will never see, have deflected attention from the narrower, if more fundamental, questions of technical feasibility. What are the chances that the particular physical characteristics of the northern environment will prevent, or modify the success of a pipeline, and more particularly, the presently envisaged gas pipelines?

To answer this question we must know the characteristics of the environment – the way the surface of the earth works in these regions. We can learn a lot from the successes and failures, and the everyday difficulties of other engineering projects: buildings, roads, airports, sewers, etc. Then the particular features and conditions which a pipeline itself imposes on its surroundings must be examined, and finally, predictions must be made about what will happen during construction, and during the life, of a northern pipeline.

1.4 WHERE DOES 'THE NORTH' BEGIN?

Our environment is a composite of immensely varied physical elements, of climate, weather, soils, vegetation, topography and of the rocks themselves. These elements can change gradually or abruptly over long distances; they do not usually change independently, nor do they change in unison. It is thus rather simplistic to do, as geographers sometimes do in elementary texts, and try to divide the earth's surface into more or less uniform regions. It is better to set a limit for an area of study by reference to some single characteristic. If one wishes to delineate the area having specifically 'northern' or cold region characteristics so far as the construction of large pipelines is concerned, it seems to be roughly that area (Fig. 1.1) north of a line where the mean annual temperature is 0°C (by mean annual temperature is meant the mean of the temperatures measured at regular intervals through the year).

This does not mean that crossing this line involves sudden new dangers, nor that all terrain within the area poses problems of equal magnitude. Nor are the problems necessarily greater than those of some warmer regions where pipeline construction is difficult – peat bogs or the earthquake-prone regions of California, are examples. However, the 0°C annual isotherm delineates conveniently a cold region – 'the North' – even though it can justifiably be argued, that in some contexts 'cold' regions lie also to the south of this. Although 0°C is the freezing point of water too much significance should not be attached to this temperature. The phenomena of freezing and thawing is indeed, important; but equally so is the fact that, in soil, there is *not* a single freezing point of 0°C. Water in soils behaves in curious ways.

1.5 THE FREEZING OF SOILS

There are a number of situations in which water does not freeze at 0°C but only at lower temperatures. One of these is, of course, when salt or other impurities are present. The freezing point of sea water is lower than 0°C, usually about -2°C with the temperature being lower as the salinity increases. The freezing point also varies if the water is in very small droplets (in clouds for example), or if the water is confined in very small spaces such as within the small pores of soils. Thus, depending on the smallness of the pores, there is water present, together with ice, in frozen soils, even at temperatures several degrees below 0°C. This is of great significance to the geotechnical engineer. Our 'everyday' experience of materials is quite often misleading, but this is a particularly interesting example. The curious behaviour of water in soils is fairly easily described, and can have the most serious consequences for a pipeline.

A sample of naturally-frozen soil frequently shows ice in the form of layers or lenses, which may be less than a millimetre or, more rarely, many centimetres thick (Pl. 1). If the sample is thawed, a layer of water will appear and lie on top of the saturated sediment. This helps explain the phenomenon of *frost heave*. But even in countries with cold climates, where people associate frost heave with jammed gates, displaced stairways, ruptured water mains and cracked or uneven road surfaces, there is often little understanding of the mechanisms involved. When a fine-grained porous material (such as soil) freezes, there is a movement of water from adjacent parts to the

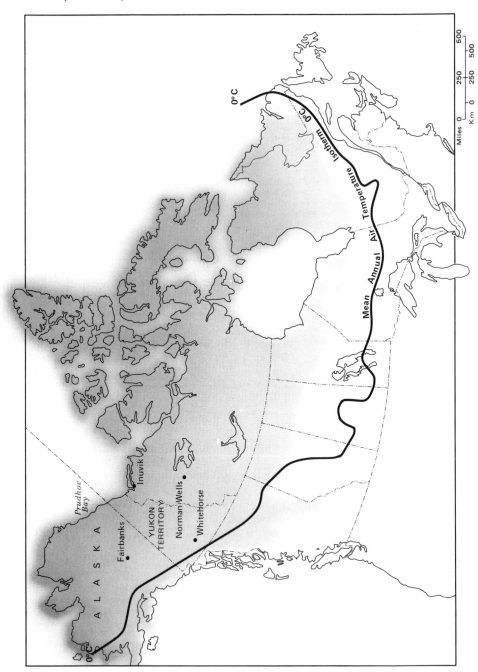

Fig. 1.1 Canada and Alaska, showing approximate areas of continuous permafrost (heavy shading) and discontinuous or scattered permafrost bodies. Also shown is the isotherm for mean annual air temperature of 0°C. (Based on Stearns 1966 and Péwé 1975.)

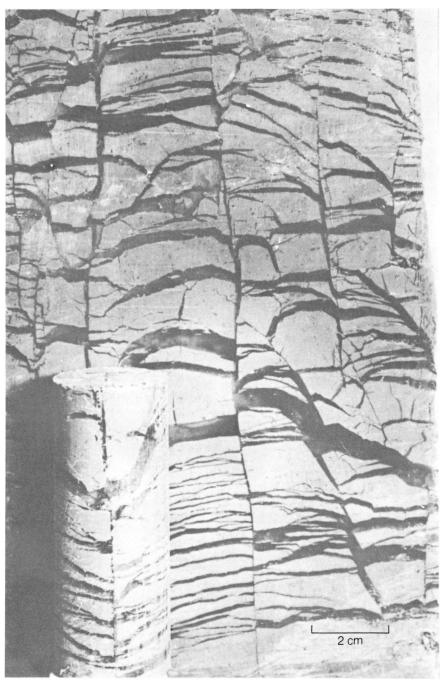

Plate 1 Frozen clay showing typical ice layers (or 'lenses'). This ice is mainly from water which is drawn to the freezing zone, and is the cause of the expansion constituting frost heave. The ice layers vary greatly in size and form, from soil to soil. (National Research Council, Division of Building Research, Ottawa.)

point where the freezing is occurring. The belief used to be widespread that breaking up of roads by frost action (a problem even in a climate as mild as that of the British Isles), was similar to the shattering of a full bottle on freezing. But the 9 per cent volume increase of ice relative to the water frozen is quite insufficient to explain observed heaving of the ground surface by as much as 0.5 m in the course of a winter – an amount quite often measured in cold areas. Clearly, the volume of the soil is augmented, by accumulation of the additional water, as ice, in the freezing soil. The phenomenon of the attraction by freezing soil of adjacent water, only became generally recognised in the 1930s. As early as 1765 a Swedish writer commented on the layers of clear blue ice he found in soil, but the hardness of frozen ground usually prevented such direct observation.[13]

1.6 PERMAFROST

In parts of the northern United States, and southern Canada, frost penetrates a metre or more into the ground every winter and the ground is much disturbed and weakened when the thaw occurs in the spring. Engineering procedures in highway and building construction routinely involve attention to these circumstances. At the same time the warm summers result in mean annual temperatures many degrees above 0°C. Why therefore should the 0°C mean annual temperature be selected as a boundary for the region of our interest? It is because a mean annual temperature of 0°C has special significance in relation to the distribution of permafrost. The word 'permafrost' implies ground which remains frozen, year in, year out. It is an emotive word and it is important to understand clearly what permafrost is, so that the connection with the 0°C isotherm can be clarified. Although permafrost (being frozen soil or rock) is such a firm material it has negative connotations. Essentially this is because permafrost is not permanent: often, under changing natural or man-induced conditions, it thaws. Excess water from the ice inclusions, coupled with the ponding of water due to underlying still-frozen material, leads to loss of strength of the soil. Huge areas of naturally-settling *thermokarst* land occur in various parts of the world, where the permafrost ice masses are thawing. The name is derived from the analogy with the karst of Yugoslavia and elsewhere, in which solution of limestone produces much subsidence. Man-made structures are subject to foundation disturbance and serious deformations if thawing occurs beneath them.

The state of being 'perennially frozen' – the definition of permafrost is ground that remains frozen right through the summer, and thus through the succeeding winter – is more likely somewhat below, than right at, the ground surface. Regardless of where we are on the earth, 'summer' and 'winter' are discernible as temperature change within a near-surface soil layer, but below about 15 m (the figure varies with the type of soil and other factors) there is no measurable temperature fluctuation through the year (Fig. 1.2). Below this depth the steady temperature is, of course, the mean annual temperature of the ground (Fig. 1.2). If it is lower than 0°C, the material is permafrost. Nearer the surface, temperatures rise above 0°C for some part of the year during which time that soil is of course, thawed. The layer above the permafrost is the *active* layer.

Thus 'permafrost' is simply material where the temperature is below 0°C for more than a year. The condition may persist for tens of thousands of years, but a near-surface layer may also thaw in two or three years. The isotherm on the map (Fig. 1.1) does not exactly delineate the areal extent of permafrost, because the mean annual temperature shown is based on *air* temperature (the only measurement available) which is not exactly

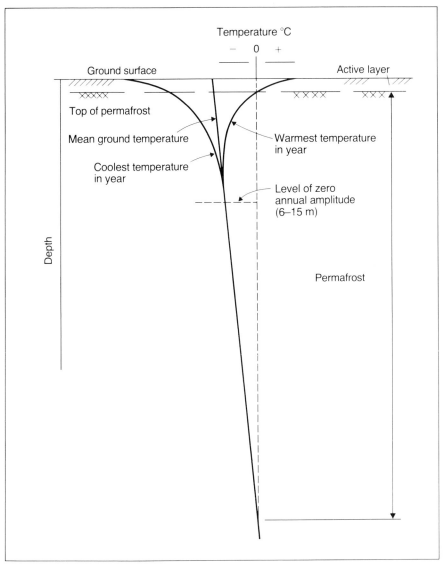

Fig. 1.2 Diagram illustrating temperatures experienced in the ground in a region with permafrost. (Modified after Brown 1970.)

the same as that of the ground. The mean annual temperature of the ground increases very slowly with depth, due to the heat of the earth's interior. But towards the surface it is determined by the above-ground climate, and the nature of the heat energy transfer processes through the surface of the ground. The mean annual ground temperature is normally different, and usually warmer by up to several degrees, than that of the air. Furthermore, the difference is by no means constant, but varies with the particular nature of the surface, whether grass covered, or forest, or asphalt pavement, etc. The relationships are very complicated; but this explains why permafrost is frequently discontinuous, that is, found in scattered bodies, or present over a wide area but interrupted by 'openings' (Fig. 1.3). .

Differences at the surface make some ground colder. If ground temperatures are around 0°C anyway, then the colder patches, but not the warmer patches, are frozen. The greater part of mainland northern Canada has such temperature conditions, this is described as the discontinuous permafrost zone. Clearly if the condition at the ground surface is changed, permafrost may be formed or thaw.

The two topics introduced in this chapter, the particular behaviour of soils on freezing, and the thermal relations of permafrost, are the basic elements of nature's challenge to the northern pipelines.

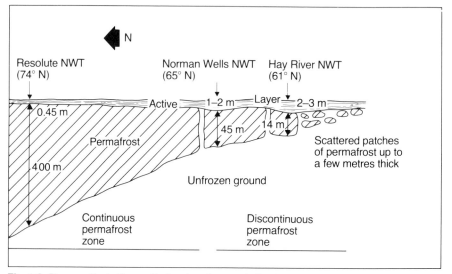

Fig. 1.3 Diagram illustrating vertical and north to south distribution of permafrost. Highly idealised profile based on information for Canada. (Modified after Brown 1970.)

NOTES:

1. Federal Power Commission 1977, Ch. I, p. 1.
2. A brief history of pipelines is given in a small, easily read book by Wynyard (1967). Details of recent developments are found in several technical encyclopedias, e.g. Considine, 1976, 1977.
3. In this book the Système International (SI) units are generally used. An exception is when a non-SI unit has become associated with a particular attribute; it would be pedantic to refer to a 48-inch diameter pipeline as '1.492 m'. When a unit is introduced, a conversion is given if this appears useful. Some common SI units and conversions are: 1 km = 0.621 mile; 1 m = 3.281 feet; 1 cm = 0.394 inches; 0°C = 32°F; 20°C = 68°F.
4. Berger, 1977, Vol. 1, p. 14.
5. Roscow (1977).
6. The costs of developing Concorde were some $2.8 billion – shared between France and the United Kingdom. Technologically successful, it is never expected to recover its development costs. But can success be measured solely by a balance sheet? Surprisingly, little is heard of the argument often used in relation to the fundamentally more questionable expenditure on weapons and related technology – that the research 'spin-off' is an additional significant gain. The aircraft's engines, however, are used in a few electricity-generating stations for remote areas. Similar engines are also used in oil pipeline pumping stations.
7. The operation of the Inquiry is described in Gamble, 1978.

8. The Federal Energy Regulatory Commission succeeded the Federal Power Commission as the agency regulating pipelines in the U.S.A., in 1977. The F.P.C. had assumed this role, although originally established in connection with production of electrical power.
9. National Energy Board (1977), Vol. 1, Ch. 1, p. 96.
10. The Agency was established for the purpose of monitoring the project after construction has been decided upon.
11. Federal Power Commission (1977), p. 2.
12. The Canadian newspaper, the *Globe and Mail*, has carried well informed articles in its business pages. In general, however, reporting of the pipeline issues illustrates the dearth of science reporters.
13. The same writer concluded that 'such ice must stretch out the ground surface where the resistance is least'. Beskow (1935) describes this and other early references to the effects of freezing on soils; some of the striking surface formations (patterned ground and solifluction) were recognised in the seventeenth century in Scandinavia. The pushing of boulders up through the surface was mentioned in a work from 1694. The existence of permafrost in Canada and Siberia was documented by the 1850s.

CHAPTER 2

THE TERRAIN
IN COLD REGIONS

Designing the foundations of buildings, bridges, highways and other constructions requires as intimate a knowledge of the ground as of the materials comprising the structure itself. For the engineer the adjacent ground becomes a part of the foundation. This is expressed in the term *geotechnical engineering*, which is only some 30 years old. Throughout this century there has been a steady struggle to establish the importance of natural conditions such as geology, soils and climate, to engineering. Even in the last two decades pipeline engineers have turned their attention to the North, knowing only that it was different because it is colder. The stories of Texans, brought to Alaska for the building of the Trans-Alaska pipeline, have become part of Alaskan folk-lore. But a lack of knowledge of ground conditions on the part of the designers of giant constructions can have serious implications. The specific characteristics of northern terrain are the basis of the difficulties associated with large-diameter northern pipelines.

2.1 PATTERNED GROUND

Early explorers and naturalists of high mountain and polar regions and especially of the treeless tundra, drew attention to a variety of forms and patterns on the ground surface, which were unfamiliar and curious.[1] Typical of the latter is the orderly arrangement of stones and boulders in polygonal or network patterns in the soil (Pl. 2). Sometimes these stone polygons cover an area many square kilometres in extent, but more often they are restricted to smaller, flat, uniform areas devoid of all but a few mosses and lichens. The phenomenon is typical of *patterned ground* – a term which also covers a variety of hummocks and other features. These include more or less circular stone 'rings', bare, or stone-covered patches, as well as a range of patterns derived from a sorting of the stones and boulders in the soil matrix. So perfect are some of the stone rings or polygon networks, that it is not surprising that they were sometimes mistaken for human artefacts. Even today the processes of formation have not been elucidated in detail, but it is clear that certain effects of the frost heave phenomenon are essential to their formation.

Even farmers in moderately cold climates are familiar with 'growing' stones. Year after year new boulders protrude through the ground surface. Such a boulder is, perhaps, the most elementary form of patterned ground. Because of the higher thermal conductivity of rock, compared to that of porous soil, freezing first occurs immediately below the stone, which is then pushed up by an accumulation of ice. The boulder is thus frost heaved relative to the surrounding soil. During the spring, thawing first occurs

Plate 2 Stone polygons are fairly common in barren places in the North. The boulders are arranged in these patterns on the ground surface by the action of soil freezing and thawing. The bleached reindeer antler is about 1 m across. Photograph taken in the Trollheimen mountains, Norway.

around the boulder, and small stones and gravel prevent the boulder from falling back into its original position. But this is certainly not the full explanation and there are many scientific papers which discuss the unexpectedly complex phenomena which the 'growing stones' represent.

Most forms of patterned ground relate to several of the following circumstances:

(a) Frost heave of soils generally occurs in a direction approximately at right angles to the layers of ice forming in the soil. This is normally towards the source of cooling. There is also a tendency for heave to occur in the direction of least resistance.

(b) Because the thermal conductivity of soils and rocks varies with their composition, compaction, and water content, it follows that cooling (or warming) will occur at different rates. It also follows that the flows of heat take place within the ground not merely in a vertical direction but at various angles to the vertical. These angles vary substantially within the area of an individual ground pattern.

(c) The degree of frost heave is very closely related to the nature of the soil, grain size, composition and moisture conditions, which are frequently very variable.

(d) Frost heave, and the associated stresses and deformations, are not completely reversible – the elements of the soil cannot return on thaw to exactly the position they had before freeze-up.

Although patterned ground occurs in many different forms, they all represent very slow, or occasional and incremental, churning of the ground. There are numerous frost heaves occurring, simultaneously, in various directions but the movements of soil and stones will often occur in a fairly ordered or geometric fashion. The movements may be shallow, or may extend the full depth of annual freezing and thawing. This depth is called the active layer and may be 2 m or more. On excavation, patterned ground commonly shows 'cryoturbation' – a flow-like arrangement of the particles and stones, which is also characteristic of certain Pleistocene geological strata which represent the soils of earlier cold periods. This disturbance of the ground by frost action is fairly ubiquitous in the cold regions, except on heavily-forested ground, bare rock, or particularly well-drained coarse sandy, or gravelly materials. It is not possible to construct a lawn-bowling green in the North.

The significance of patterned ground lies not in the surface forms themselves, but in the great forces indicated by the movements of soils, stones and boulders. These movements occur only very slowly, the formation of patterned ground takes decades, but the forces behind them are powerful enough to affect man-made structures. Breakup of roads and building foundations, and damage to water and sewage pipes are familiar to Canadians, Scandinavians, Russians and others, even those in fairly temperate climates; thus the effects of freezing have come to dominate design and construction procedures at higher latitudes.

2.2 SOLIFLUCTION AND OTHER SOIL MOVEMENTS ON SLOPES

The effects of gravity are important in all natural processes occurring at the earth's surface, particularly in landslides, mudflows and other events which modify the form of slopes. In cold climates, the effects of freezing and thawing combine with those of gravity, to produce downslope movements of the active layer. A particular kind of movement which involves freezing and thawing, is referred to as solifluction.

Although solifluction takes various forms, it characteristically involves development of 'step' ridges or 'wavefronts' often a metre or more in height, and occurring at intervals of metres or tens of metres upslope. Viewed from afar, the surface of the sloping ground resembles small waves lapping over a sand beach, or perhaps thick sauce poured in successive spoonfuls over a pudding (Pl. 3). But commonly the solifluction lobes, terraces, fronts, are not fluid or soft. Solifluction features are, with exceptions, quite firm to walk on, even during the wet spring period.

There is no longer any doubt that these features are the result of downslope movement. The advancing fronts overrun such vegetation as lies downslope (the phenomenon is typical of the treeless, tundra areas) and the buried organic matter dates back thousands of years (Fig. 2.1). The movement is imperceptible because it amounts to only a few centimetres per year.

The term solifluction does not include landslides or earthflows (both terms imply a more rapid process), but refers only to those soil downslope movements caused by certain effects of the freezing and thawing of the soil. It produces surface forms unique to the cold regions. Landslides, mudflows, etc. are also caused in cold regions by other factors such as snow meltwater and impeded drainage. These other features may be even more abundant than solifluction.

The stability analysis of slopes by design engineers and others, is based on the concept of forces within the slope, essentially arising because of weight, which are

Plate 3 The characteristic forms of the ground surface where solifluction is taking place, arise in association with a slow movement downslope of a layer of soil 1–2 m deep. The photograph was taken at a location about 30 km from that of Plate 2. There is much variety in the effects of freezing on soils, even over much shorter distances than this.

resisted by the strength of the soil or rock material comprising the slope. The disturbing forces may be increased by steepening the slope, by adding extra mass, and in other ways, while the strength may be decreased by some change in the nature of condition of the material. In either case, a sudden displacement of soil downslope – a 'failure', landslide, mudflow, soil slip – occurs when the forces tending to cause movement exceed those tending to resist movement (the strength).

By their nature, many of the slow solifluction movements are not amenable to this kind of analysis. There is no identifiable instant of failure, and the slopes can, in fact, support additional weight without such failure occurring. Standard engineering analyses may conclude that the slope is extremely stable – the forces tending to move material downslope are substantially less than the strength or resistance of the materials. The slow protracted movements which nevertheless occur are then described as creep. This term is used differently in engineering and the earth sciences. It is best regarded as very slow, fairly continuous deformation of a material, which occurs at stresses

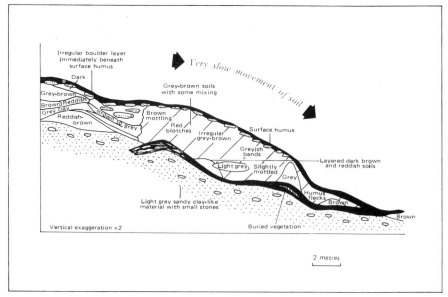

Fig. 2.1 Profile in slope affected by solifluction. A dark organic layer is apparent, which represents vegetation buried by slow downslope movement of the soil above it. The longest-covered part is almost certainly thousands of years old. (From Williams 1957.)

(stress = force per unit area)[2] well below those required to cause a flowing, sliding, or 'rupture' – that is, at stresses below those equalling the strength, as commonly defined. Creep phenomena can be extremely important in frozen soils, and we suppose, also occur in association with thawing and subsequent drainage.

There are other ways in which the process of freezing and thawing reduces the strength of soils. Much of the strength of fine-grained soils lies in the forces interacting between particle surfaces to hold them together, by 'cohesion'. The disruption of the soil structure by expansions of ice, separates particles and destroys much of this cohesion. 'Internal friction' is the second source of soil strength. This is the effect of friction between particles, which must be overcome if the soil is to deform. The friction increases with the pressures of the particles bearing upon each other. The buoyancy effect due to water in the pores counteracts these pressures and reduces the friction between the particles. In thawing soils, abundant water arises from melting ice inclusions, as well as from ground surface drainage and snow meltwater. The majority of landslides in the world occur in association with relatively wet conditions.[3] Thus, it is not surprising that in cold regions, where water can accumulate as snow or ice, and particularly where there is poor surface drainage, landsliding and mudflows of various kinds are common.

In an extremely detailed study in Swedish Lapland, Professor Anders Rapp[4] established that, quite contrary to the view often propagated in textbooks, solifluction was not the dominant process of slope denudation (in the geomorphological sense) of the cold 'periglacial'[5] regions. Although solifluction is characteristic of such regions, the total movement of soil and rock by 'ordinary' landslides and related phenomena was far greater.

As in more temperate regions, the incidence of all downslope soil movements will vary greatly depending on local factors and situations. The Mackenzie and other rivers in Canada have innumerable slides initiated by river erosion of the banks, and associated with summer thawing of a deepened active layer. Although they pose practical difficulties for engineers, their cause is quite well understood. From a scientific point of view the mystery surrounding some of the solifluction phenomena is much greater. Small annual movements which decrease below the surface and do not extend much deeper than a metre, do not suggest great danger. But they have, on occasion, been responsible for deformation of railway tracks, highway and other foundations, and slow deformation of the foundations of a pipeline could be serious. There is also the problem of identifying soils prone to these movements. Experimental measurements have shown quite 'normal' slopes, without visible characteristics of solifluction, to be susceptible.

2.3 ICE WEDGE POLYGONS, PINGOES AND PALS

All the characteristic features of the cold regions relate in some way to the presence of ice. The phenomenon of frost heave, with the formation of ice layers or lenses in the soil is fundamental in the development of patterned ground and solifluction. But there are other, even more remarkable ice bodies to be found in strata of frozen ground which are sufficiently thick. Perhaps the most widespread are vertical ice wedges in the permafrost. At the base of the active layer, they are a metre or so wide; they narrow downwards to a depth of 2–3 m. In plan the wedges are about 10 m in length, and are arranged in such a way that from the air a polygonal network is seen on the soil surface (Pl. 4). Of course, this is just another example of patterned ground, but what is unusual is that the disturbance caused by the ice wedges extends into the permafrost. The pattern is due to cracking of the permafrost caused by contraction on cooling during intense winter cold. It is similar to what happens when mud dries, except that the surface polygons are commonly 15 m in diameter. Each winter the cracks re-open, and water, water vapour or snow, enter and slightly enlarge the wedge. Careful measurements have shown that the wedges enlarge by only a few millimetres each year. But through tens of years, centuries, or even millennia, networks of ice wedges have covered enormous areas of the permafrost regions, frequently extending as far as the eye can see from an aircraft. During the summer, there is warming of the permafrost and a re-expansion of the ground which is pressed against the ice wedges and forced to rise around them resulting in the characteristic, banked ditch appearance.

A less common feature, although even more striking, is the pingo which is an ice-cored hill some tens of metres high (Pl. 5). In Canada and Alaska they are most common in wet, flat areas and sometimes resemble a small, dead volcano (the top being eroded and gullied). The interior of the pingo is a large body of ice. Painstaking research by Dr J. Ross Mackay, over many years, has established that the core of ice develops by increments of as much as 25 cm per year, after being initially formed during development of permafrost.[6] The fact that *developing* permafrost is involved restricts the distribution. But even well within the permafrost regions, there are transient unfrozen sections. One such section lies under the entire length of the Mackenzie river. The constantly flowing water means the bed is maintained above 0°C so there can be no permafrost beneath it. But the river shifts course through centuries, and abandoned channels become lakes, which in time shoal over. Ultimately they will freeze through. When ice extends to the bed for much of the year, permafrost forms and spreads below the shallowing lake. This has happened repeatedly to give the conditions for the formation of the pingoes of the Mackenzie Delta. Although the pingoes most commonly found are of this type, and in low-lying old lake beds, there is a second type which is

Plate 4 The tesselated surface of permafrost terrain seen from several hundred metres height. The network of lines (each line is a metre or so wide on the ground) marks the location of underlying ice bodies which extend downwards in a wedge form into the ground. Each winter cracks form along these lines, by contraction of the ground, and the wedges enlarge slightly by addition of ice.

found at the foot of slopes. In these the ice core is presumably nourished by ground water moving down the slopes to accumulate under pressure. In the Mackenzie Delta type, water is apparently extruded from the freezing sediments, while the ice accumulation also involves, to some extent, the same thermodynamic processes as frost heave.[7]

Superficially similar, but smaller, features, are 'pals' (pronounced 'pulse', from the Swedish: palsa, plural: palsar), found near the southern limits of the permafrost areas in the 'discontinuous' zone. These are mounds up to 10 m in height, although usually much smaller, and with a core of permafrost. The core usually has a lot of ice. Pals are quite abundant where there is a substantial surface layer of peat. Presumably the mineral soil content is sufficient to promote frost heave, as pure peat does not heave. It appears that frost heave initiates elevation of the pals' surface; the surface is then more exposed, and

Plate 5 A pingo. These hills are formed by the expansion of an ice-filled core, and grow, sometimes by 25 cm each year, to 30–50 m height (note figures on top of pingo). Photo from Yukon, Canada, nr. Dempster Highway.

is thus increasingly cooled during the winter. In the summer the top of the pals dries and vegetation then insulates it from the heat of the sun. This explains the isolated permafrost body which gives the pals its form.

2.4 OTHER GROUND ICE AND THERMOKARST

As the thickness of permafrost increases northwards, in response to lower mean ground temperatures, the permafrost is more likely to include large masses of ice, up to 10 m thick and quite distinct from ice wedges. Such ice masses may be abundant in low-lying or wet areas but are often not revealed by any ground surface expression. Their origins are various. Some are the result of a prolonged freezing process, with migration of water to the growing lens to result in a frost heaving that must have continued for years. Others are simply surface accumulations of river ice, or perhaps snow, which have become covered by sediments, or by mudflows or landslides. Sometimes the buried ice may be glacier ice but this is not generally the case. Rarely it seems, have both the soil conditions and thermal regime been such as to preserve such ice since the glaciers retreated at the end of the last glacial period.

Little was known about subsurface ice until the widespread drilling activities of petroleum companies in the last 20 years, and even now exploratory drilling is normally necessary to elucidate subsurface conditions.

When these ice masses thaw, their former presence is revealed by subsidences and sink-holes at the surface. When there is widespread thawing of permafrost containing such ice, the effect is to produce characteristic thermokarst. The classic regions for thermokarst are in Siberia, where large areas covered by small ponds (called *alas*) and lakes represent the early stages of a progressive thawing of permafrost containing numerous ice wedges. A succession of changes follow, as the ponds tend to coalesce and then shoal over (Fig. 2.2). These changes are often, in a geological sense, quite rapid, sometimes taking place in a human lifetime. They represent a fairly extreme example of how climatic change which is barely perceptible without careful temperature measurements, may cause rapid environmental change. Of course, equal in importance to climatic change is the fact that most permafrost is already quite near to its melting point.

2.5 GROUND THERMAL REGIME

There are many places in northern Canada and Alaska where the terrain illustrates the rapidity of change over a large area associated with thermokarst formation. French (1976) has described an extreme example in Banks Island, within the Canadian Arctic, where there is a 'badlands' topography, distinguished by rapid erosion of slopes caused by a combination of thawing and slumping. As much as 8 m of material is lost from the surface of the slopes in a single year.

A more intimate knowledge of the landscape shows us that the thawing of permafrost has an even more widespread and fundamental significance. When permafrost thaws it produces, in addition to the settlement phenomena described, a layer of weak material, the loss of strength being due to the excess water from melting ice. The top of the frozen ground often defines the shearing surface for sliding soil masses such as those common on the banks of the Mackenzie and other rivers. Such a surface, after exposure, is the site of renewed thaw.

Permafrost is an effective barrier to the drainage of water. It can happen that thawing occurs in the permafrost (deepening the active layer) in such a way that the hidden permafrost surface takes the form of a ditch or trough. This may in turn act as a channel for subsurface water flow. As the water introduces more heat, thawing progresses, surface settlement begins, and a new, eroding, surface stream or gully appears. Firebreaks, vehicle tracks, and other disturbances which destroy the surface vegetation, often initiate such a sequence, which can ultimately lead to substantial modifications of surface drainage and stream patterns.

A key issue for the understanding of these phenomena are the factors which control the growth and decay of the permafrost or, more especially, of its more surficial or peripheral parts. Undoubtedly climate is one cause. But climate as we normally understand it, has a certain uniformity and does not change significantly over small distances. In fact, it is the *microclimate* with which we are concerned. Microclimate concerns the nature of the earth's surface as much as it concerns the atmosphere. The thermal conditions of the upper layer of the ground depend on climate (as we generally understand the word), and on the nature of the energy transfers across the boundary

Fig. 2.2 Thermokarst: sequence of development and disappearance of alases.
1 Original lowland surface with ice wedges
2 Effect of increasing depth of active layer
3 Initial thermokarst stage
4 A young alas
5 Mature alas
6 Old alas
7 Stage of possible pingo formation
8 Thawed pingo
(Modified after Czudek and Demek 1970)

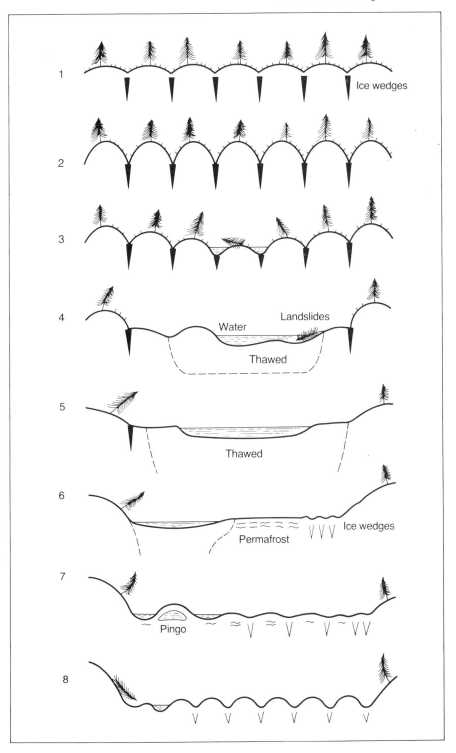

between the air and the ground. Features such as river banks, vehicle tracks, narrow gaps in the trees, are each associated with particular microclimates. If, for example, the sun's radiation falls on snow, a greater quantity is reflected back than if the surface is green vegetation. Such heat as arrives at the ground surface may be utilised in part in the evaporation of water; on the other hand, condensation of water liberates heat. Heat energy is always arriving at the ground surface and always leaving, in diverse ways (Fig. 2.3). Conduction of heat upwards or downwards in the soil is one of these, and controls the temperature distribution in the soil.

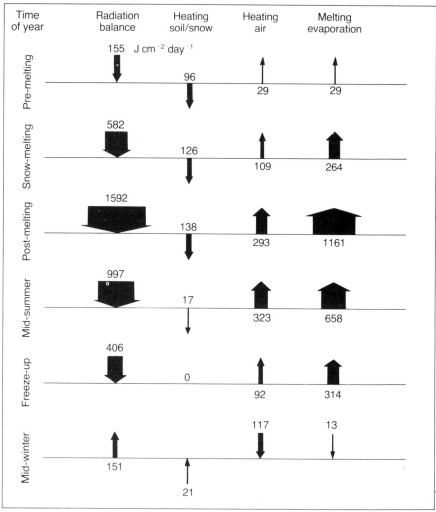

Fig. 2.3 Energy arrives at and leaves the earth's surface in various ways. This diagram gives values for six occasions during the year, at a site in Alaska. The arrows show the direction of the energy flow and whether the energy path is in the ground, or above the surface (horizontal line). The radiation balance refers to the receipt of energy mainly from solar radiation, less the radiation of energy outwards from the earth's surface. (Modified after Weller and Holmgren 1974.)

The energy arriving at the ground surface can never just 'disappear' and thus we arrive at the important concept of heat balance. In whatever ways heat energy is arriving at or leaving this surface, at any given time the two sum totals, for 'arriving', and 'leaving', must be equal. A corollary of this is that if any of the processes, evaporation for example, occurs at a different rate (suppose the ground surface has become dry) then there are immediate, compensating adjustments in the rates of some other of the processes. Any natural surface is subjected to such general processes as radiation exchanges, evaporation and condensation, and arrival or departure of heat by conduction. The details obviously vary, but they are always many, complex, and constantly changing. It is usually impossible to predict by analysis of the heat balance, with any degree of precision the effect of a particular change of the ground surface on the thermal conditions of the soil. In other words, it is not usually possible to say precisely how the quantity of heat leaving or entering the soil (Fig. 2.3) will vary. Normally, in temperate climates, such effects are of interest only to the agronomist or horticulturist, who recognises from experience warmer or colder soil situations. But when the temperature of the ground is within a degree or two of freezing, it is clear that even very small changes to the ground surface can lead to modifications of heat exchange, and a freezing or thawing of the soil which can have dramatic consequences.[8] The discontinuous permafrost is discontinuous in time, as well as in space.

If importance is attached to short term changes in the nature of the ground surface, it must not be forgotten that the atmospheric climate is also constantly changing. There is abundant evidence for changes in average air temperatures, or average precipitation in historical time – changes that are reflected in famines, migrations and other events.

The soil erosion of the 'dirty thirties' in the Prairies of Canada, the much more recent droughts in parts of Africa, and the 'minor' advance of glaciers in the eighteenth century are all examples of especially rapid activity by the natural (or semi-natural) terrain-forming processes. Each of these situations was, to some extent, due to the fact that the climate itself had changed as well as to changes in the nature of the ground surface. Large areas of the north, as we have defined it, are on the threshold of similar disturbances associated specifically with freezing and thawing and the ground thermal regime. Man's involvement is both to provoke, and to suffer from, such disturbance.

NOTES

1. A book which gives a comprehensive description of the nature of the northern terrain, with some emphasis on North American conditions is French (1976). A more expensive, richly illustrated work is that by Washburn (1978), and there is a simpler text by Embleton and King (1976).
2. Most people think they know what force is, but find it hard to define. One definition is 'that which changes the state of rest or motion of matter. . . .' (Weast, 1978.) Force is measured in Newtons and the force acting on a unit area (e.g. a square metre) is called a stress. Force and stress have direction ('which way it pushes'). A pressure can also be expressed in N m^{-2}, but applies equally in all directions (consider the pressure of a gas or liquid).
3. Although elementary textbooks emphasise the obvious effects of abundant water, like many 'obvious' matters, the truth is more complex. It is actually the pressure of the water in the soil which is of particular importance. It relates to bouyancy. Pressure in a fluid can be thought of as a tendency to 'spread out' (consider water leaking from a hose, or air from a tyre). The pressure of the water in the pores of the soil (the 'porewater pressure') depends on various conditions, such as depth, and slope. When the soil is saturated, the porewater pressure represents expansive forces, and by tending to push the soil particles apart, reduces the friction at their points of contact. The attraction of the particles one for another, the *cohesion*, is also reduced when the particles are less closely packed. Thus the strength of the soil is reduced and a landslide or other rupture or failure becomes more likely.

CHAPTER 3

A BRIEF HISTORY OF GEOTECHNICAL ACTIVITIES AND ASSOCIATED SCIENTIFIC RESEARCH IN THE NORTH

Explorers, fur traders and missionaries had travelled in the permafrost regions of Alaska and Canada for centuries, but it was not until the end of the nineteenth century that there was a significant transfer to the North of activities associated with industrialised America. The native peoples, that is, those whose presence predated the arrival of Europeans in the North American arctic and sub-arctic regions, traditionally lived by hunting and fishing. Even after the establishment of trading posts and settlements, the natives were not caught up in the explosive pace of development and change occurring elsewhere and which had its origin in the industrial revolution. But, particularly in this century, improvements in transportation and the expansive tendencies of industry and capital made changes in the North as inevitable as changes in the colonies of empires. Since the Second World War, human activity in the North has usually been dominated by, and always influenced by, industrialisation. The history of these changes is well-documented, but much of it is current knowledge for it occurred within living memory. From even a brief review it is apparent that certain lessons of that history are being overlooked in the current, accelerated, development.

Northern development is aimed at the exploitation of natural resources which are transported elsewhere for utilisation. This circumstance, coupled with the special characteristics of the northern terrain, mean that development has been, and remains, largely a matter of geotechnical science and engineering.

3.1 THE PASSIVE, OR PRE-TECHNOLOGICAL, APPROACH

It is not surprising that the phenomena described in the previous chapter are important for human activity. We have only to think of what such forces and movements of the soil would do to the foundations of our homes in temperate climates, to see that substantial adaptations will be necessary under northern conditions. The native peoples traditionally lived in relatively simple structures which were moved or vacated at frequent intervals. It was this mobility rather than any technical solution, that protected their dwellings and surroundings from the effects of their own activities. All subsequent northerners, in their desire for more permanent quarters and amenities, have incidentally set in motion a continuing series of detrimental changes, often associated with thawing of ground and sometimes with its freezing.

That even the simpler activities of man initiate such changes, is illustrated by the example of a path worn, in 10 days, by the feet of an Eskimo's chained dog.[1] The wearing down of the vegetation modified the exchanges of energy through the ground surface, and led to thawing of a layer of the permafrost with a subsequent 38 cm of settlement of the ground in the next three years. The process would have continued except that the vegetation slowly recovered. When vegetation covers disturbed ground, permafrost forms again. Frost heave is likely, and ultimately re-elevation of the surface.

All human communities in their striving to improve their conditions risk inadvertently modifying their environment. In this respect it is important to remember that because so much of the ground in the North is at temperatures near 0°C a relatively minor event may result in freezing or thawing with concomitant radical changes in behaviour and nature.

Even in southern Canada, the northern United States, Scandinavia and elsewhere, the spring break-up of roads can still lead to traffic restrictions; prevention of frost, damage to building foundations, highways, and services such as water mains, is a major design and cost factor. Early settlements by the white newcomers in the North were beset with similar problems, but at a different level. Technical solutions to frost problems, were, 50 or 100 years ago, rather elementary in the fairly temperate and inhabited regions. In more extreme situations and in more distant parts, the problems were largely tackled by improvisation, or ignored.

Plate 6 Old house in Dawson City, Yukon, Canada. Thawing of permafrost, particularly below the warmer, central parts of buildings, commonly led to subsidence. When a building falls into disuse, renewed freezing and frost heave produces further distortion.

Buildings were fortunately constructed of timber which is better able to withstand the stresses resulting from uneven displacement of foundations. In the gold rush towns of the Yukon, buildings were similar in design and construction to those of many southern Canadian towns. Their foundations were generally massive timbers ('mudsills') laid directly on the surface of the ground. Some of the buildings still stand, distorted and leaning, in Dawson and elsewhere, but many succumbed to heaving or settling of the ground beneath them (Pl. 6).

It is interesting to consider the case of the town of Aklavik, a meeting place for fur trappers and an administrative centre for the North-West Territories of Canada for many years. At first impression the town might appear to be favourably situated, on the flat expanse of a silty-clay terrace in the Mackenzie Delta. The site is of course underlain by permafrost. However, both the permafrost and the active layer when frozen, contain many times more ice than sediment because of the abundant water and highly frost-heaving nature of the soils. Early buildings sometimes had conventional concrete basement walls which became cracked and displaced when ground ice melted, while later, larger buildings were built on timber posts (piles), although even these did not prevent serious distortions. Modern sanitation or water supply never became possible, because buried pipes were frequently ruptured by enormous heaves during the annual freezing of the upper layer of the ground. The annual thaw of this active layer reduced roads to a quagmire, impassable by wheeled traffic (Pl. 7).

Plate 7 Photograph taken in 1957, showing summer condition of roads in Aklavik, N.W.T. Flat wet ground on permafrost contained large amounts of ice in winter and became a quagmire in summer, preventing the use of wheeled vehicles.

In Aklavik, as the years passed, the depth of the active layer increased below buildings, trails, roads and other constructions where the ground surface was changed from the natural state. Unfortunately, most man-induced changes, for example, the removal of vegetation, or the escape of heat from a foundation, increase the thickness of the active layer. In contrast, a decrease in the thickness might have been beneficial.

Although the Aklavik town site is particularly unfortunate the circumstances are repeated, to some extent, everywhere in the North. The manifestations of the freezing or thawing of ground, and the effects this has on vegetation, stability of slopes and the quantity of sediments in rivers, vary greatly. In some ways harsh cold winters present more stable and predictable conditions than the short, sometimes warm, summers which are periods of continuous thawing. Trafficability is greatly increased during winter when the ground is frozen hard. Transportation in connection with exploration for oil and gas, or construction, is concentrated in the winter months. The dangers of stagnant water, improper sewage disposal, and pollution are greater in the summer months.

Public health considerations, and the difficulties of achieving acceptable standards of performance for larger buildings, led the government of Canada to plan to vacate Aklavik, and to initiate, in 1954, the building of the replacement town of Inuvik. The location and building of a town of this size, in virgin terrain, is an unusual opportunity. The prime requirements for a site were that the soils should not be susceptible to frost heave, which meant a site on relatively coarse-grained soils and with sufficient elevational relief to be well-drained. Soils composed exclusively of grain of sand size and larger do not show frost heave. Inuvik is on high ground, yet adjacent to a navigable river. There is little ground ice, and gravel is at hand for foundation and road improvement. The town is an example of the successful application of the elementary principles of site selection for northern conditions, coupled with the introduction of techniques evolved specifically for northern construction.[2]

3.2 POST-WAR NORTHERN DEVELOPMENT AND THE GEOTECHNICAL APPROACH UP TO 1960

During the Second World War military activities, the construction of airfields, highways (notably the Alaska Highway through the Yukon), and pipelines such as the Canol (Norman Wells to Whitehorse, and Fairbanks, Alaska, a 5-inch oil line), and the enlargement of several towns, resulted in a more organised approach to northern construction problems than previously. The establishment of Inuvik and the rejection of any further development of the long-extant Aklavik illustrates clearly the direction that was to be followed, even without the consideration of military imperatives.

The first English language book on permafrost engineering (and in a narrow sense, the only book) appeared in 1943, sponsored by the U.S. Army Corps of Engineers.[3] Mostly however, the engineering techniques were developed as problems presented themselves, rather than being learnt from books or manuals. Larger vehicles and bigger aircraft in the post-war period meant that settlements and cave-ins of highways or airstrip surfaces became more dangerous. It became apparent that, in permafrost areas, the effects of melting ice were not limited to the 'break-up' and mud of a short-spring period, but could involve unexpected collapses as large sub-surface permafrost ice bodies slowly melted away. Clearly, it was the bare surfaces of the highways and airstrips themselves, which were initiating progressive failure through the years. Covering these surfaces with a metre or so of gravel was quite successful in some cases in limiting the thaw phenomena. The gravels were not, of course, prone to frost heave.

Highway construction involved consideration of the adjacent terrain. Damage or removal of vegetation, particularly in cuts, initiated a thawing which, for example, in the early Alaska highway often led to mudflows, washouts, or other events affecting the stability of the highway itself. Gradually the importance of route selection, avoidance of ice-rich and poorly drained areas, the desirability of coarse-grained soils and sediments, and the importance of the natural vegetation cover, were realised.

Plate 8 Thermokarst pond. This particular pond developed on the site of a vehicle parking place beside the Dempster highway, Yukon, Canada. The loss of the natural vegetation cover resulted in a warming of the ground over several years, which was sufficient to thaw several metres of permafrost. The permafrost contained much ice, and a pond developed which itself led to further thaw. Natural thermokarst, formed without the intervention of man, gives abundant ponds like this one.

The technique of the insulating and supporting gravel mat was extended to small building foundations. For larger buildings, pile construction became general. Following the practice of lake dwellers in many parts of the world, buildings were constructed on top of wooden or steel piles inserted into the permafrost. The technique is also widely used in temperate climates where the weight of modern buildings may be too great for the upper, weaker soil layers. To prevent the passage of heat from a building into the permafrost, a well ventilated crawl space is left by allowing the piles to project a metre or more out of the ground. The building shields the ground from solar radiation and care is taken to preserve the natural moss or peat layer as far as possible. Nevertheless, some early pile-supported buildings settled irregularly because the piles became unanchored as the ground thawed. In some cases, this retreat downwards of the permafrost was probably due to a general warming of the ground associated with climatic change: in the period 1880–1940 air temperature records suggest a warming of the climate occurred over wide regions. Even imperceptible changes can result in the advance or retreat of glaciers, or changes in the extent of permafrost, especially if the ice masses are already within a degree or so of the melting point.[4] Heaving of the active layer may lift piles if these are insufficiently anchored in permafrost, and piles are now inserted into the permafrost to at least twice the depth of the ultimately expected active layer.

Conventional foundations, suitably protected against seasonal frost heave effects are used where permafrost is absent. There are large areas in the southern part of the discontinuous permafrost zone in Canada and Alaska, where the permafrost is so scattered or restricted in area that it may normally be avoided. So long as the highway or building foundations and their immediate surroundings do not overlie permafrost there is no long-term risk of unacceptable thaw and settlement. The problem, of course, is to identify the permafrost-free ground. Intimate knowledge of local conditions, of soils, drainage, plant life, coupled with previous observations from drill holes and excavations, facilitates reasonable predictions. But so complex are the natural energy exchanges, and so many the variations in vegetation, soils, topography, etc. that such an approach is not infallible. Frequent confirmatory drilling and sampling is necessary. Even then, a 2 m wide, 10 m long body of permafrost is easily missed.

Permafrost intrudes on many other activities. The primitive methods of sewage and household waste disposal still practised in small communities present problems. Excavation of frozen ground for burial is difficult, and the accumulation of wastes themselves may initiate thawing and serious settlement of the ground. During the summer months surface water and mud are easily polluted. The utilidor, heavily insulated, boxed-in pipes used to carry sewage, or water (some provision for heating may be necessary), was introduced in larger communities in the fifties.

Other attributes of modern civilisation have been successfully introduced to the northern environment: the oil tank farm; the electric transmission mast; mineral crushing plants and other appurtenances of mining; railways; river and lake dams; and various constructions of modern military activity; radar and other communications installations. Appropriate design modifications were often necessary. The traditional and almost exclusively northern activities of hunting and fishing and a semi-nomadic existence were not to remain – and, clearly, technological innovations in engineering and construction procedures have hastened the changes. The innovations were not, for the most part, based on sophisticated scientific study and research. They involved fairly simple ideas such as the use of large quantities of non-frost-heaving gravels for foundations, improved piles, whilst other special anchoring devices were devised for towers. Very largely, innovations were based on common-sense modifications of normal engineering practices. To this day, many civil engineers have no formal training in northern geotechnical procedures, and only a few universities have specialist courses on the cold regions and their natural characteristics. Experience gained on the job suffices only for the simpler engineering challenges.

That this period, up to the late 1950s, may fairly be regarded as the heroic age of northern geotechnical engineering is illustrated by the Russian situation. There, northern towns such as Noril'sk and Irkutsk had, even by that time, wide, paved streets, and large masonry buildings not unlike some larger Scandinavian urban centres. Yet these Russian towns are underlain by permafrost. In fact, the rigid buildings suffer from cracking, floor displacements, etc. to an extent that might well be considered intolerable by North American standards. But these buildings symbolise the larger scale of Russian activities. As befits, perhaps, the greater number of persons inhabiting the Russian north, by 1930 a substantial permafrost research institute had already been established, to be followed by several other study centres so that by the 1950s many hundreds of persons were engaged in research. Codes or 'norms' to be used in northern design and construction were developed for standardising many procedures. There were, as we shall see, some highly significant Russian scientific advances. For all that, Russian engineering and construction procedures are not much more advanced than those in North America. It would be misleading to suggest that Russia has an overall technological lead.

It is the extent rather than the sophistication of the Russian involvement that has become apparent. Technological advances have been made in the design of water, sewage and heating services to buildings, piles and anchors for foundations, and other items such as the heated concrete conduits used to carry several individual pipes. The latter show significant refinements over the simple utilidors constructed mainly of wood and more widely known in Canada.

By 1970 a number of large dams for hydroelectric generation had been constructed on the permafrost in Russia (a more limited experience had been gained in North America). A gas pipeline was in operation between Tas Tumus and Yakutsk involving 400 km of relatively small diameter, 500 mm (19.7 inches), and also 325 mm (12.8 inches) pipe. Some 60 per cent of the length is elevated on piles, 8 per cent lies on the ground surface, and the remainder is buried.

Pressures in the pipe were some 3 500 kiloPascals.[5] But in 1971 the Director of the Permafrost Institute of the Soviet Academy of Sciences said that large diameter gas pipelines in permafrost would present greater problems than those already constructed and had not been studied.[6] By 1978 there was still no gas pipeline of large diameter buried in frozen ground. At the beginning of the 1970s there was a sudden surge of interest in the transport of gas and oil by pipelines in northern North America. There was a widespread assumption that this would be by large diameter (42 inch, 48 inch or larger) pipes, with gas pressures of over 7 000 kPa. This is about 35 times the pressure in a motor car tyre. Clearly the dangers inherent in such pressures require consideration.

3.3 THE SCIENTIFIC APPROACH

From 1960 progress in northern construction technology has been marked by an increasing amount of scientific research and analysis, particularly in studies of what actually happens when soils or other porous materials freeze, and studies of the exchange of energy through the ground surface. An understanding of these phenomena is a requirement for design engineers whose structures must be adapted to the way the earth's surfaces behave. An understanding of the soil freezing process involves the study of phenomena which are at the very heart of classical physics and thermodynamics, modified only by the special situation represented by the finely porous nature of soil. Energy exchange at the ground surface, and its variation with the nature of the vegetation and other features, is fundamental to microclimatology and also involves thermodynamics. Local variations in the distribution of permafrost are, of course, microclimatological in origin.

Traditionally permafrost and geotechnical studies were the responsibility of geologists, whilst construction activities were the concern of soils' engineers with little special training for northern conditions. Since neither microclimatology, nor the thermodynamics of freezing and thawing were normally part of the professional experience of either group, many of the key explanations have been made by more academic scientists, research workers in institutes or universities, who were able to pursue quite basic theoretical studies.

3.4 WHAT HAPPENS WHEN SOILS FREEZE?

In the middle 1950s the Russian scientist, Nersessova, using calorimeters of the kind available in schools, found that cooling, or warming, of *frozen* soils, involved the addition or extraction of amounts of heat much greater than expected.[7] Of course, freezing or thawing of water involves transfer of a large quantity of heat (331 Joules must be removed to freeze one gram) compared to that required to warm water (4.2 Joules per gram, for each degree of temperature). But Nersessova's observations involved soils which were not at 0°C. In fact, we now know that frozen soils down to several degrees below 0°C, contain both water and ice (Fig. 3.1). The greater part of the world's frozen ground is at such temperatures, and thus is not the solid and inert substance it was once thought to be. Indeed the freezing, or thawing process occurs over a range of temperature, and not just at 0°C. The presence of the unfrozen water, at temperatures below 0°C is an example of the well-known phenomenon of freezing-point depression.

The existence of water, as well as ice, in frozen ground is of great importance. It is essential for an engineer, for example, to understand the properties of the materials with which he works. Difficulties arise when materials do not behave as predicted. Only when the presence of the water was understood, and its quantity to some extent established, was it possible to calculate accurately the thermal properties of freezing soils, and thus how much they might freeze or thaw in a particular situation. The proportions of ice and water change with temperature, and with pressure, and thus affect the way in which frozen soil may deform, rupture, or flow. Furthermore, if there is liquid water in the permafrost, this could presumably move, just as water passes through other permeable rock materials. Biologically too, the liquid water might present a habitat for certain bacteria and other organisms, and perhaps raise new questions of soil contamination. Knowledge of the presence of water in frozen soils and of its behaviour was to have particular significance in designing the large diameter gas pipelines; this is described in Chapter 5.

It is curious that the coexistence of ice and water in soils was not recognised until Nersessova's work, and that the pressure states of the ice and water phases were similarly overlooked. These pressures are the origin of the stresses generated by frost heave. That the freezing point of water changes with pressure, or with the addition of soluble material, or when the water is in very small droplets, was known to the Victorian physicists. To some extent these phenomena are also a matter of observation. Already, by the middle of the last century, some extremely important and fundamental equations and relationships had been discovered, and these enable us to understand scientifically, a host of related everyday phenomena. It is the application of these 'elementary' principles, that constitutes most of the recent advances in our understanding of freezing soils.

The first historical event of relevance was the work of the French mathematician Laplace,[8] who in 1805 presented an analysis of capillarity – the phenomenon whereby water moves up fine tubes ('capillary' tubes), or into porous objects (absorbent tissue, sponge, etc.). Soils being finely porous, are capillary media. Laplace demonstrated that in a capillary, one phase, say water, is at a different pressure than another adjacent phase, say air. It is important to be able to visualise pressure, in the way Laplace was able to do but which scientists two hundred years earlier were not. Pressure is a property we associate with expansion or displacement, although this only occurs if there is an adjacent region of lower pressure, into which mass can move. Water squirts from a leaky bucket due to the increase in pressure caused by the weight of the water, and continues

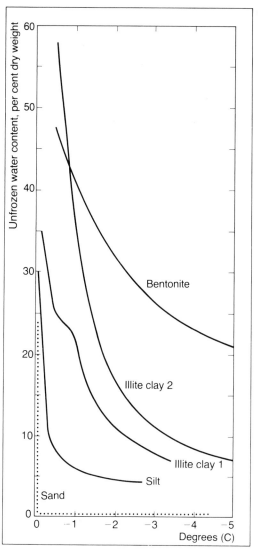

Fig. 3.1 *Water* contents of frozen soils. Water remains unfrozen as a result of capillary, osmotic and surface adsorption effects, but is progressively replaced by ice at lower temperatures. Thus the water contents depends on temperature and type of soil.

until the increment of pressure is dissipated. Laplace's special contribution was the concept that different pressures are *maintained* across a small interface, such as that in a capillary. It is the lower pressure of the water at the capillary meniscus (Fig. 3.2) which results in water moving up a capillary tube, because water from below flows upwards towards the lower pressure. It soon stops, but the difference in pressure across the meniscus persists. The lower pressure results from forces originating with the water molecules in the meniscus, and those in the surface of the tube. Essentially the forces act equally in all directions. But the forces of the molecules in the meniscus are not balanced

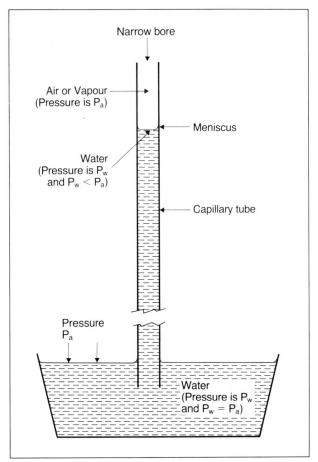

Fig. 3.2 Water moves up a capillary tube because of the lower pressure (compared to that of the air) adjacent to the meniscus. The water rises to a certain equilibrium height, which depends on the diameter of the tube. If this, for example, is one-thousandth of a cm the height is approximately 3 m.

in that the meniscus is a discontinuity, and small in area. The lack of balance determines the lower pressure, and increases the smaller the radius.

Other scientists continued to study capillarity, but the event of significance for our understanding of freezing soils was the publication of a paper, in 1834, by a French railroad engineer, Benoît-Pierre-Emile Clapeyron.[9] Clapeyron demonstrated how the pressure of water vapour existing above ('in equilibrium with') a body of water, varies with temperature. Again, there appears the important concept of two, coexisting phases (water and vapour) with *different* pressures. Water vapour is not, of course, associated only with boiling. Humidity results from vapour in air. The vapour pressure is not the same as the pressure of the atmosphere: it is less. Boiling point temperature is an exception, for boiling occurs precisely because the vapour pressure then overcomes atmospheric pressure. Of course whilst Clapeyron's findings applied to any size of body of water, in a consideration of soil the microscopic nature of the pores must also be taken into account.

The significance of Clapeyron's work was highlighted by that of Rudolf Clausius,[10] a Prussian scientist who rejected then current views of the nature of heat, but stressed its convertibility to work (that is, force applied through distance). Clausius' studies were important in the development of the subject of thermodynamics as a whole, but of most interest to us is his extension of Clapeyron's findings. What is now known as the 'Clausius-Clapeyron' relationship, refers to the relation of the *pressures* in two phases existing side by side (for example a solid and its liquid) to the temperature. This relationship applies in a number of situations including that of the water in frozen soils which is coexisting with the ice (Fig. 3.3). Thus for a range of freezing temperatures (Fig. 3.4), the Clausius-Clapeyron equation tells us about the difference between the pressure that occurs in the ice, and that which occurs in the water in the freezing soil.[11] These aspects are considered further in the analysis of the origin of 'frost heave' pressures – which are essentially the pressures of the *ice* (Ch. 5). The ice can increase in volume except when the pressure of the overlying soil is as great or greater.

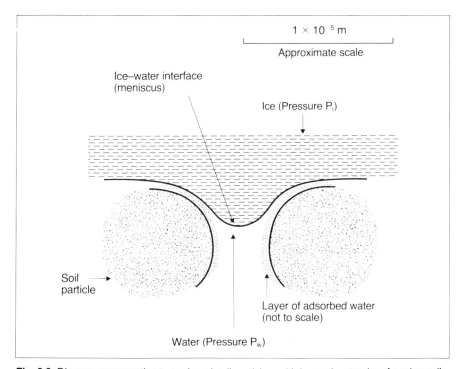

Fig. 3.3 Diagram representing two mineral soil particles, with ice and water, in a freezing soil.

By combining the Laplace capillary relationship with Clausius-Clapeyron, we can see the extent to which the sizes of pores in soils are responsible for the freezing occurring over a range of temperatures (Fig. 3.1), and for the development of particular pressures. It was Lord Kelvin's work in particular later in the nineteenth century which showed, in the 'Kelvin' relation,[12] how size (of capillary), temperature, and the pressures of the two phases were interrelated.[13] We should also note that in fine-grained soils a significant amount of the water is so close to the particle surface (we are talking of layers a millionth of a centimetre thick) that its properties are modified by molecular

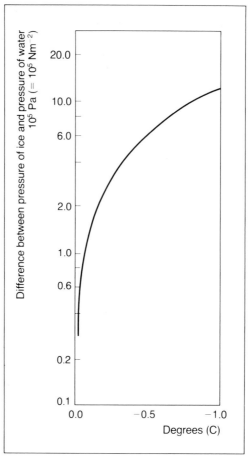

Fig. 3.4 Graph showing the difference that apparently exists between the pressure of the ice and of the water in freezing soils, at various temperatures. Somewhat similar relationships apply in other natural situations, for example, clouds.

forces from the surface itself. Although much remains unknown about the water so affected, it seems that it too behaves *as though* it had the pressures implied by the Clausius-Clapeyron equation.[14] The uniform existence of this adsorbed water film means that there is a 'path' (or rather, innumerable 'paths') of water right through the assemblage of particles making up the soil.

The fact that highly complex situations can often be described as though they were much simpler, is basic to the studies that constitute classical thermodynamics which were developed during the second half of the last century.

Josiah Willard Gibbs[15], whose name has been given to 'Gibbs free energy', contributed substantially to the understanding of the generality of natural phenomena, as expressed by the thermodynamic functions. The idea of a quantity, the free energy, controlling the tendency for transfer of mass from one state or position to another, has proved of particular use in analysis of complex situations such as those represented by for example, living organisms, or, for that matter, movements of water in soils. Simple hydrostatic pressure is not the sole cause of such movements and we do not always know the causes.

This outline is intended to show that scientific research into frozen soils is based on the application of principles which are elementary to those versed in basic science. It is only in the last 25 years, however, that this has occurred despite the fact that in meteorology, for example, these same relationships have been recognised for more than 75 years. An outstanding contribution prior to 1950 was made by the Swede, Beskow, whose still highly valued and detailed experimental study of 1935[16] directed attention to the boundary between the frozen and unfrozen soil – where the temperature is of course, essentially 0°C.

Even during the 1960s the results of fundamental scientific studies of freezing soils found little application in 'cold regions'' geotechnical engineering. The scientists' work provided, through occasional academic meetings, or by publication in journals, a muted background commentary to the trials and errors of the highway and foundation engineers, a commentary which, on occasion, served as an aid to the engineers' intuitive and empirical approaches. The immensity of the trans-Alaska, and Mackenzie Valley pipelines, however, required accurate prediction of the long-term behaviour of the soils. This produced a sudden demand in the early 1970s for applied scientific analysis. Unhappily, the process of transition from science to technology cannot always be made successfully in a short time.

3.5 WHO STUDIES FROZEN GROUND?

The development of highway networks in several northern countries had led to a shortage of the gravels and sands used as foundation materials. The shortage could, presumably, be alleviated by sounder and more economical foundation design. In Norway, a major government-funded research programme involving some ten researchers and associated staff over the period 1969–76, led to increased knowledge of the energy exchange at the highway surface, and of the thermal and other properties of freezing soils. This programme produced important scientific findings and led to a rationalisation of frost design practices in Norway.

In France too, detailed consideration has been given to highway design for areas of cold winters. In the United States, the Army Cold Regions Research and Engineering Laboratories has for many years pursued a programme of high quality scientific and technical or engineering research. A few universities have courses which deal with the geotechnical problems of cold regions specifically. Most deal with the cold regions as a natural phenomena of only academic interest.

The University of Alaska has a substantial involvement in both teaching and research, in northern geotechnical and scientific matters. Russian construction achievements have been referred to several times, and there the hundreds of people involved in research into frozen ground represent a national involvement greater than elsewhere. Much Russian research has a very strong practical or applied bias and their contributions in this respect are much more important than in the more fundamental and theoretical fields.

In Canada studies were, for many years, represented by a small group at the National Research Council. A few rather isolated scientists working mainly in universities, made important findings, which however did not greatly influence geotechnical practice. It was the conservation movement as much as the coming of the pipelines which shook the country out of its complacency and provoked an awareness of the special characteristics of the terrain, fauna and flora of the North which after all constitutes more than half the country.

3.6 CONSERVATION AND THE CONCERN FOR THE NATURAL ENVIRONMENT

In the late 1960s, popular opinion finally became alerted to the power that man, in peace as in war, had to change the face of the earth, and rallied to the preservation of the 'natural' environment. The preservation of certain areas of land in a natural or semi-natural state is not a new idea. The first national parks in the U.S.A. were established in the late nineteenth century. What distinguishes the recent surge of interest is the concern for a particular 'environment' as a coherent system. This is usually called an *ecosystem*, a term which includes both biological and physical components. Areas of concern are the consequences, for the integrity of a system, of a disturbance, even minor, of one or more elements and the less obvious effects on the populations of individual plant or animal species. Because it is a wilderness, the North has provided examples of all these different kinds of disturbances.

The passage of seismic exploration vehicles over the treeless tundra kills a narrow strip of vegetation. In the course of a few years there is usually a deepening of the active layer, and some settlement of the ground, and if the ground is sloping erosion of a gully may be initiated. The intensity of the search for oil and gas means that some areas are criss-crossed by these tracks. Such highly visible changes suggest vandalism and awaken concern that the few wilderness areas of the earth relatively unaltered by man are valuable and should be protected, as are irreplaceable artefacts, or historic buildings which also have a special non-monetary value.

The report of Mr Justice Berger's inquiry into the Mackenzie Valley pipeline proposal deals at length with such questions as the effect on the populations of mammals, of certain birds, such as the rare peregrine falcon, and also of fish species which have considerable economic importance to the native peoples. His report advocates the preservation of the northern Yukon from intrusion by a pipeline. The region is of special importance because it contains the nesting sites of Arctic birds and is a caribou migration route. These animals could face substantial depletion of their numbers, or even extinction, as a result of the disturbance of their habitat. As an alternative the pipeline for gas from Prudhoe Bay could be routed along the Alaska highway, through the southern Yukon. This route is the subject of the present Alcan pipeline proposal.

It is now evident that pipeline and petroleum companies laying pipelines in remote areas, will have on occasion, to yield to considerations of environmental conservation. The force of public opinion is such that it has become politic for government to create regulations to this end. At the same time the adverse effects of pipeline construction, or the lack of it, on the lives of native or other long-term residents, must also be considered. Even when the desirability of a pipeline is established, the study and resolution of these problems is an expensive and slow process.

In discussions of pipeline issues in the popular media the terms 'environmental problems' and 'social problems' are rather common. The terms imply the desirability of preserving the *status quo*, of reducing the incidental effects of a pipeline on its surroundings to a minimum and ultimately of doing without, or at least relocating, a pipeline. Naturally such attitudes (and the legal requirements which have followed) present difficulties to the pipeline constructors, who are wont to consider those who espouse such views – the 'environmentalists' – as a greater problem than the environment itself.

These attitudes are to some extent suggested by names such as 'Environment Protection Board',[17] a regrettably short-lived Canadian review agency supported by the petroleum producers themselves. The even more emotive but general term, *environmental impact assessment* – describes a study to establish the effects of proposed construction.[18] Unfortunately, considering the 'environment' as a passive element, open to all manner of abuse, is to miss the main point of current 'environmental' or 'ecological' studies. These are based on the interdependence of one element upon another, and of processes or events upon one another. Just as the pipeline will disturb the environment, so can the environment disturb the pipeline.

In fact, the first question should be: what effects will the environment have on the pipeline and what problems will result? These problems have to be met by design, construction and maintenance procedures. It is the success or failure of these procedures which will in large measure determine the magnitude of disturbance to the natural environment. Temporary failures of the pipeline, and unusual remedial procedures, emergency excavations and repairs, probably represent the most intense, although localised, environmental disturbance that could occur.

The construction of major transport routes, such as railways, canals, roads and highways, has, through history, often been a question of overcoming the natural difficulties and hazards of the terrain. Engineering has to be successfully adapted to the demands of the particular environment. In the case of the northern pipelines, while construction is regarded as an engineering 'challenge', there has been a tacit assumption that the technological difficulties are surmountable. Indeed, they almost certainly are, if not now, then at some time in the future. But, as the next chapter will show, the technological problems for the pipeline constructor, arising from the northern environment, are diverse and severe. These problems must be overcome before financial success is possible. Upon their solution also depends the satisfactory conservation of the environment.

NOTES

1. Mackay, J. R. (1970).
2. The book by Brown (1970) is the best, easily read general survey of building and construction technology in the North.
3. The book by Simeon W. Muller (1943) is regarded as a minor classic, and has an important place in the history of northern development.
4. To speak of the 'warming of the climate' (or worse, the 'amelioration') as a *cause* for such effects, is actually rather unscientific. The diminution of glaciers is just as likely to be due to a decrease in precipitation, as to a rise in temperature. A decrease in the extent of permafrost represents a warming of the ground, and thus of a warmer climate below the ground surface. But as has been stressed, the temperature of the ground depends on many factors – the temperature of the *air* above is only one possible factor.
5. Although pressure, like stress, is force per unit area, and can be expressed in $N\ m^{-2}$, this quantity is often called the Pascal with reference to pressure ($1\ N\ m^{-2} = 1$ Pa). Car tyre pressures are about 200 kiloPascals.
6. On the occasion of the official visit to Siberia of the Canadian Minister of Indian and Northern Affairs (Slipchenko 1972).
7. Nersessova's work was originally published in the proceedings of the Academy of Sciences (Institut Merzlotovedenia im V.A. Obrucheva 1954–57). Although many Russian scientific and technical works relating to frozen ground are translated into English by the U.S. Army Cold Regions Research and Engineering Laboratories, the National Research Council of Canada and by others, the only widely available book in English is Tsytovich (1975).
8. Pierre Simon Laplace, French, born 1749, died 1827. A theoretical mathematician, and astronomer. He has been described sometimes as second only to Newton.
9. B-P-E Clapeyron, born Paris 1799, died 1864. As a civil engineer, he designed steam locomotives, bridges, etc. Although a professor at L'Ecole des Ponts et Chaussées from 1844, he completed his perhaps most famous paper (on vapour pressure and temperature) in 1834 when he was active in the design of railways and locomotives.

10. Rudolf Clausius, born Russia 1822, died Bonn, 1888. A somewhat chauvinistic physicist, his correspondence about priority of discovery, with allies and foes alike, in other countries, was typical of the period.

11. There are in fact several forms of the Clausius-Clapeyron equation. This form shows the relationship between temperature, and the pressures of ice and water:

$$\frac{P_i - P_w}{\Delta T} = \frac{l}{TV_w}$$

T = temperature, K

ΔT = change of freezing point

V_w = specific volume of water ($m^3 kg^{-1}$)

l = latent heat of fusion, J kg^{-1}

For example in the case of a freezing soil without a load, the pressure on the ice, P_i, is atmospheric. A sub-atmospheric pressure, or suction is then developed in the water, of magnitude P_w.

The following form will be recognised as the 'normal' equation for the dependence of the freezing point of ice on pressure, where the pressure is the same on the ice and on the water:

$$\frac{dP}{dT} = \frac{l}{T}\frac{\Delta V}{}$$

dP = change of pressure

dT = depression of freezing point

ΔV = change of volume on freezing

12. The name is given to equations of the type:

$$R T \ln \frac{P}{P_o} = \frac{2 \gamma V}{r}$$

R = universal gas constant

T = temperature

P = vapour pressure above curved surface

γ = surface tension

r = radius of curvature of surface

P_o = 'normal' vapour pressure (over plane surface)

V = molar volume

The equation shows how the pressure of the vapour above a curved surface (of a meniscus, or droplet), depends on the radius of the surface, and on the temperature. With certain qualifications, similar equations are applied to ice and water when in small spaces (soil pores).

13. The English physicist, Poynting (1852–1914) perceived the manner in which the freezing point of ice would change when the pressures of ice and water were different, and devised an experiment with a block of ice on sand to demonstrate it (Poynting 1881a). He published a paper in the Philosophical Magazine but subsequently withdrew his conclusions, in a humble letter to the editors, when he could not measure the expected temperatures (Poynting 1881b). Such letters, indeed such self-criticism, seem curious by today's standards. The irony is that Poynting correctly described the equation for freezing of water in soils. His experiment failed because he did not carefully consider the size of the pores of sand which are relatively large! Some 65 years passed before the relevance of the equation was again realised.

14. Because the particles of a clay are so small ($<2\ 10^{-6}$ m) they have, collectively, a large surface area. There can be 800 m^2 of surface in a gram of clay. Even if the layer of water around the particles, which is affected by the adsorption forces, is only a millionth of a centimetre thick, it follows that there can be more such water in the soil than clay mineral! The existence of the 'modified' water is important in explaining the general behaviour of clays.

15. Josiah Willard Gibbs is also distinguished by the fact that from 1871–1880 he was Professor of Mathematical Physics at Yale, but received no salary. He had private means but an offer of paid employment from Johns Hopkins University forced Yale to rectify the situation. J. W. Gibbs was also, initially, an engineer.

16. Beskow (1935).

17. The Board under the chairmanship of Carson Templeton, a distinguished engineer, made important submissions to the Berger Inquiry.

18. Such assessments and their role are reviewed in Mitchell and Turkheim (1977).

CHAPTER 4

THE TRANS-ALASKA OIL PIPELINE

In only a decade, since the end of the 1960s, the oil and gas industry has come to dominate economic considerations of the North of the American continent. Yet, in the last decade, the only wells producing in northern Canada were those at Norman Wells. These supplied only a local market and had done so for nearly half a century. In 1977, however, the larger diameter oil pipeline which crosses Alaska to Valdez was completed, the oil at Valdez being transferred to tankers for California. This pipeline was the first to transport northern petroleum to southern markets. Its construction demonstrated the impact of such a concentration of capital and manpower, and indicates the effect future pipeline projects can be expected to have on northern Canada.[1]

Oil men are prone to use superlatives, but the claim that the Trans-Alaska pipeline would be the largest single engineering project in history undertaken by private industry was, however, true. The eight-fold cost escalation, from the $900 million predicted in 1968, to over $7 billion, ensured that it was also the most costly.[2] 'Single engineering project' refers to the fact that not until the 1 300 km of pipeline was *complete*, and the oil flowed could the undertaking be a success. The economic and social effects of failure on the State of Alaska would have been profound, and the financial ramifications would have been international.

There has been much discussion of the delays in the building of the pipeline, but an outsider might well be surprised at the short time which elapsed between conception and completion.

4.1 THE FIRST BIG PIPELINE ON PERMAFROST

In 1968 geological investigations established the existence of a large oil field in the vicinity of Prudhoe Bay on the north coast of Alaska (Fig. 4.1). The significance of this field was that it was a source of oil for the United States 'south of 49'; a significance which received timely emphasis in the 'energy crisis' of 1973. By February 1969, Atlantic Richfield, British Petroleum and Humble Oil had already announced plans for the giant pipeline project.

As tankers can only reach Prudhoe Bay during the ice-free season, a few months each year, the oil is transported 800 miles overland by pipeline, to Valdez on the southern coast, before being shipped to U.S. west coast ports. The 48-inch diameter steel pipe constituting the Trans-Alaska, or Alyeska pipeline (as it became known), has a wall thickness of only about $\frac{1}{2}$ inch. Its contents are at a temperature of about

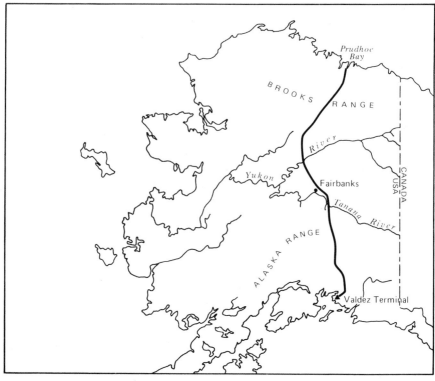

Fig. 4.1 Map of Alaska showing main physical features and location of Alyeska oil pipeline. Permafrost is essentially continuous in the Brooks range and further north and to the west. Discontinuous, or scattered bodies of permafrost occur throughout the rest of the state except for a small permafrost-free southern coastal strip. (Permafrost information from Péwé, 1975.)

60°–65°C. The oil leaves the well at approximately 80°C and if allowed to cool to air temperature, would become viscous and waxy.

No other oil pipeline in the North approaches this one in either diameter or length. The Bechtel Corporation, the contractor responsible for the Alyeska line, was however, also responsible for the trans-alpine pipeline running from near Trieste, through the Alps to West Germany. Although subjected to rugged winters, that route did not cross permafrost.

The Alyeska line traverses a great variety of forest, swamp, mountain, and tundra terrain, and represents a remarkable engineering achievement. However, the eight-fold escalation in cost is worth further consideration, particularly as the sums involved are so large.[3] Some 1.2×10^6 barrels per day of oil reach Valdez currently, that is, nearly 200 million litres, and this is expected to rise to some 2.0×10^6 barrels per day. At full capacity the flow will approximately equal the amount imported to the U.S. from the OPEC countries, and will be about 15 per cent of total U.S. production.[4] Since the flow of oil commenced in 1977, there have been only short interruptions caused by a technical mishap in the operation of a valve, by the collision of a bulldozer with the pipe, and by a case of deliberate damage to the pipe.

The first hearings into the pipeline proposals were held by regulatory agencies in 1969. Four years of assessment and controversy followed before a permit for construction was issued in 1973, and construction started only in 1975. The reasons for, and significance of, these protracted preliminaries can largely be ascribed to the challenge presented by the northern terrain. Coupled with that was an increasing public awareness of, and concern for, the effects of such extensive works on their surroundings.

4.2 PERMAFROST AND EARTHQUAKES

Remarkable as it now seems, the earliest proposals were to construct the pipeline in a conventional manner, burying it in the ground for virtually the entire length. In fact, the most basic problem was that the warm pipeline might thaw the underlying permafrost. Wherever the permafrost contained quantities of ice in addition to that within the soil pores, settlement or subsidence would inevitably follow thawing. At one planning stage serious consideration was given to using a chilled pipeline: cooling the oil so that the pipe could be buried without thawing the permafrost. This proposal was rejected because of the effect of low temperatures on the oil, which would not flow satisfactorily. The interaction of a chilled pipeline with the soil, which was to present problems for subsequently-proposed gas pipelines, was apparently not foreseen.

Probably three-quarters of the proposed route overlay permafrost, and at least half of this was estimated to contain ice, which on melting would cause settlement. The effect of the warm pipeline was such that up to 10 m of soil could be expected to thaw in the first year.[5] The amount would vary, of course, depending on the pre-existing ground temperatures and the type of soil.

Near Prudhoe Bay the permafrost is some 600 m thick. Southward it becomes generally thinner although the thickness is very variable. Thawing around a buried, warm pipeline would progress downward, although at a decreasing rate, for many years. The degree of consequent settlement, or subsidence, would depend on the amount of 'excess' ice in the thawed layer, but quite often there would be several metres displacement. Obviously, such effects left unchecked would cause great disruption of the pipeline.

The second problem which had to be considered when designing the pipeline was the risk of earthquakes. The southernmost section of the pipeline passes through the edge of the zone within which the disastrous Anchorage earthquake of 1964 occurred. In addition there are several rock faults in the interior of the region and earthquakes have been recorded within this century. In fact, all but the northern third of the route was potentially at risk. However, pipelines already exist in earthquake-prone areas, so the problem was not new. But in this case it was additional to those problems imposed by the cold climate.

Another area of concern was the possible effects of construction operations on the terrain. These 'environmental' considerations extended further than a desire for conservation and avoidance of disruption of natural ecosystems and landscapes. In some locations it could be demonstrated that while damage to the natural environment had relatively minor immediate effects, processes of erosion, or disturbance of natural drainage would be initiated that could eventually threaten the pipeline itself. The remedial measures then necessary would be costly and, at the same time, cause further disturbance. The consequences of an oil spill from a ruptured pipeline and the delay in operation would be more disastrous than the original environmental disturbance. While terrain disturbance was seen by environmentalists as undesirable, the process set in motion by construction operations would also affect the future operation of the pipeline. Thus the pipeline companies had a vested interest in the interaction of the pipeline with its environment.

4.3 TERRAIN CONDITIONS AND SITE INVESTIGATIONS

The Alyeska Pipeline Service Company ('Alyeska') was set up by the petroleum companies owning the system, to design and construct the pipeline. Alyeska initiated a programme of research which included major field studies to select the best route, and subsequently to ascertain more precisely the conditions along the route which was eventually chosen. Some 5 000 bore holes, to ascertain soil and ground conditions, were put down in the initial stages, at a cost of about $10 000 each, or $50 million in total.[6] This may seem a very detailed investigation, but it averages out at only 6 holes per mile which is hardly excessive considering the variation of vegetation, soils, and topography that occur in that distance.

Although the Alyeska route traverses such a wide variety of terrain, it was the ice in the permafrost which, directly or indirectly, presented the major and unique problems. Ice in permafrost may be almost invisible, or it may be present in layers several metres thick. Only in coarse-grained soils, gravels, and sandy materials, particularly those that are fairly dry, will there be so little ice that thawing will not result in some settlement.

Those parts of southern Alaska where permafrost is absent posed no new problems. A freezing and thawing restricted to the winter months never extends more than 2 or 3 m into the ground. A pipeline may safely be laid in such ground, if appropriate design precautions are taken.

Locating and investigating the permafrost is, however, essential. There is in Alaska, as elsewhere, a broad belt of country where permafrost occurs in a discontinuous, or 'scattered' fashion (Fig. 4.1). This is the region, described in Chapter 1, where the temperature of the ground is *near* to 0°C. If the temperature is less than 0°C at a depth greater than that at which there is any summer warming (so that the temperature is the same, year in, year out), the ground is obviously permafrost (Fig. 1.2). The temperature of the ground can, however, vary by a degree or more within an area of a few hundred square metres or less.

These variations in ground temperature result in a complex way from the different values of the ground surface energy exchange components, discussed in Chapter 2. Normally, the only way to identify these colder patches of ground is to bore a hole to see if it is frozen or to be well acquainted with the particular locality. For example, in mid-Canadian 'discontinuous' regions (in Saskatchewan and Manitoba), permafrost may be absent under well-drained mature forests, present in low-lying peaty ground, while being absent below pools or lakes (Fig. 4.2). In the iron-ore mining area of Schefferville, Quebec, for example, permafrost occurs only on treeless hilltops, and is generally absent in the wooded valleys.

Because the general characteristics of terrain vary, as one moves across a landmass, it is difficult to establish rules for the localisation of permafrost bodies. This was indeed

Fig. 4.2 Diagram illustrating the distribution of discontinuous permafrost in relation to local topographic and drainage features, microrelief and vegetation. The diagram is highly idealised, but illustrates the relation of permafrost to the nature of the surface. The surface modifies the ground temperatures by its effect on the energy exchange between the ground and the outside. (Modified from Brown 1968.)

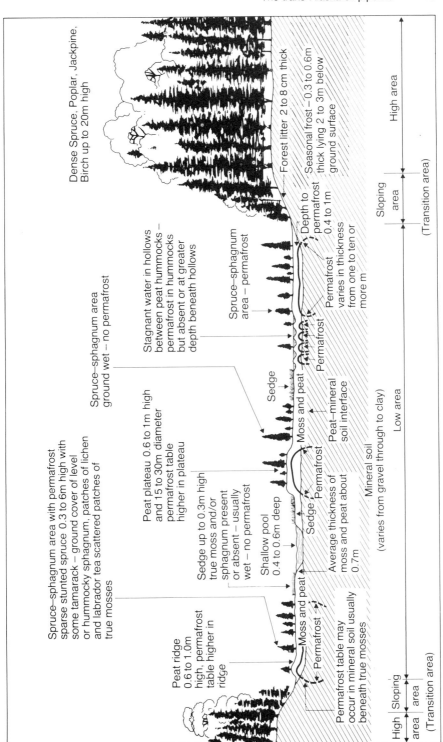

Dense Spruce, Poplar, Jackpine, Birch up to 20m high

Forest litter 2 to 8 cm thick

Seasonal frost – 0.3 to 0.6m thick lying 2 to 3m below ground surface

High area

Sloping area

(Transition area)

Spruce–sphagnum area ground wet – no permafrost

Stagnant water in hollows between peat hummocks – permafrost in hummocks but absent or at greater depth beneath hollows

Spruce–sphagnum area – permafrost

Depth to permafrost 0.4 to 1m

Permafrost varies in thickness from one to ten or more m

Permafrost

Permafrost

Sedge

Moss and peat

Peat–mineral soil interface

Low area

Mineral soil

(varies from gravel through to clay)

Spruce–sphagnum area with permafrost sparse stunted spruce 0.3 to 6m high with some tamarack – ground cover of level or hummocky sphagnum, patches of lichen and labrador tea scattered patches of true mosses

Peat plateau 0.6 to 1m high and 15 to 30m diameter permafrost table higher in plateau

Sedge up to 0.3m high true moss and/or sphagnum present or absent – usually wet – no permafrost

Shallow pool 0.4 to 0.6m deep

Sedge

Permafrost

Average thickness of moss and peat about 0.7m

Moss and peat

Peat ridge 0.6 to 1.0m high, permafrost table higher in ridge

Moss and peat

Permafrost

Permafrost table may occur in mineral soil usually beneath true mosses

High | Sloping area | area

(Transition area)

the case for the 800 miles to be traversed by the Alyeska pipeline, and the soil engineers had, therefore, to develop their own criteria. There are certain patterned ground features, such as pals, pingos, and ice-wedge polygons, which are usually reliable indicators of the presence of permafrost.[7] In addition thermokarst terrain, with its depressions and ponds caused by thawing ground ice, is easily recognisable. But these indicators are not always present in a 'discontinuous' zone. And sometimes, as in the case of 'ice wedge' polygons, the surface configuration persists long after the permafrost has thawed. 'Fossil' polygons occur for example, in New Jersey, East Anglia in the United Kingdom, and scattered throughout central Europe.

The engineers planning the pipeline were thus faced with the immense task of locating permafrost bodies, which in a discontinuous zone may be only a few metres or tens of metres across. Even when the permafrost was delineated, the presence or absence of excess ice had to be established. 'Excess ice' refers to ice in excess of the normal volume of soil pores, such that if the ice thaws there will be settlement of the ground. The term 'ground ice' is used more or less synonymously, usually for ice bodies that are conspicuous in exposures. Ground ice is revealed by the surface patterns of ice-wedge polygons; excess ice is also more likely in flat, wet areas, but predicting its presence and amount remains a difficult problem for northern geotechnical engineering.[8] Various remote sensing devices, some of which can be used from aircraft, have been developed for investigation of sub-surface conditions. These include methods for measuring the electrical resistance of the ground, or for measuring the dielectric constant. Both properties have substantially different values in a frozen soil and soil that has thawed. Electromagnetic wave transmissions or radar devices, which transmit impulses into the ground from a surface source, or alternatively from probes inserted into the ground, are quite widely used. These devices have other applications in locating bedrock, ore bodies or other buried strata. But their use for locating ground ice is still in an experimental stage, and it may be difficult to distinguish a boundary between frozen and unfrozen materials from other, lithological changes. Normally, the existence of permafrost or ground ice has to be confirmed by taking a fairly large number of control borings, and by direct observations. Most of the preliminary investigations for the Alyeska route were based on the drill hole programme and correlation with surface mapping of soil and ground surface features. It was boreholes put down for installation of the pipeline foundations which provided the final check. On occasion the findings at this late stage resulted in local modifications to the location of the pipeline foundation.

4.4 THE PIPELINE AND HYDROLOGICAL CONDITIONS

The necessity of preventing the thawing of ground ice by heat from the pipeline was obvious, but there were other potential problems to be considered. Natural drainage is disturbed by the presence of a pipeline and its associated structures, and this can ultimately threaten the pipeline itself.

If the pipeline is buried, and covered by an earth berm or mound, it is a potential barrier and can divert streams or sub-surface drainage paths. During the winter months large mounds of ice build up, particularly where culverts and other drainage structures are required. Such mounds of ice, known as 'icings' or by the German word *aufeis*, are fairly common under natural conditions, for example where springs issue from the ground. They are more common in association with highways and associated structures. Aufeis also presents problems for pipelines supported above the ground, for the supporting structures can hinder the movement of surface waters, or of floating ice.

Particularly where there are shallow streams passing under the pipe, the build-up of ice can threaten the support structures themselves.

There are other far reaching effects which involve erosion and sedimentation. Construction procedures result in accelerated melting of ground ice because the stripping of surface vegetation and soil disturbs the heat balance. Normally, this will be followed by increased sediment loads in rivers and streams. Most of this sediment originates from the erosion of soils in new water courses, developed in association with irregular patterns of thawing and subsidence. A further major source of sediment is the increase of landsliding into streams and rivers.

Increased sediment loads may have a serious effect on fish populations. They may also be a problem if the waters are utilised for hydro-electric production. The sediments may accumulate behind storage dams reducing their effectiveness, and they may also damage turbines through which the water is passed.

Apart from possible earthquakes, water, whether liquid or ice, appears to be the essential element in the geotechnical difficulties of this northern pipeline. There are large areas with high concentrations of ground ice; many stream and river crossings; and generally poor drainage. It is surprising, therefore, to find that the greater part of the Alyeska route has low rainfall and snowfall. Indeed the amount of precipitation for the northern one-third of the route is so low that by some definitions it is semi-arid. The North Slope for example (the name refers to the land on the north side of the Brooks Range) has on average a mere 18 cm of water equivalent (it mostly falls as snow) per year. Yet this is the region where ground ice and thermokarst ponds are especially abundant. The central part of Alaska has less than 38 cm. Seventeen centimetres is, in fact, similar to precipitation in parts of the Sahara, the Gobi desert, or the great Victoria desert in Australia. By comparison, the average annual rainfall in London is 61 cm and in New York 109 cm.

This 'dryness' is, in fact, characteristic of much of the high Arctic. In the Canadian Arctic islands there are areas where the dryness is indeed immediately obvious in the terrain, which is sometimes, with some justification, referred to as 'Arctic desert'. Such a description would seem most inappropriate to anyone familiar with Alaska's streams and swamps, buried ice, in both the north or south, and the widespread forests of the central and southern parts.[9]

The question, why such moist conditions are associated with such low precipitation, can only be answered in general terms: the relatively low temperatures of the air reduce evaporation, but, more important, the presence of permafrost hinders the downward infiltration of water. This is a clear example of how inappropriate a reliance on precipitation figures as a climatic characteristic may be. The effects of climate on a natural environment are far too complex to be successfully characterised by such elementary parameters as annual precipitation. There was one direct consequence of the low snowfall: it was sometimes difficult to prepare winter roads by the standard procedures of snow compaction.

4.5 SOLUTIONS TO THE PROBLEMS

Preliminary studies for the pipeline concerned the rate of thaw that would occur around a buried pipeline on account of its temperature.[4] Subsequently the problem became that of finding designs that would not cause unacceptable thawing. The necessary mathematical techniques for prediction were available. But while the design engineer can refer to a book of tables for the appropriate values for the thermal behaviour or properties of most materials, very little information is available for freezing soils. No two soils are exactly alike, of course, and freezing soils have only recently become the subject of detailed study. And in several respects freezing soils behave in ways which are still not understood. For this reason it was decided to carry out experiments on a length of hot pipeline, and trial installations were constructed for this purpose at Fairbanks, where 200 metres of pipeline were laid down.[6]

As soil surveys proceeded much more ground ice was discovered than was initially predicted. This, coupled with findings as to the amount of thaw that would occur, led gradually to the decision to build more and more of the pipeline above ground elevated on pile supports, rather than buried in the ground as first envisaged. The air passing beneath an elevated line would dissipate most of the heat from the pipe and greatly reducing the thawing of the permafrost. Furthermore, the pile supports, or 'vertical support members' (known as VSMs) could be designed to permit lateral movements of the pipe as it expanded or contracted with temperature changes (Fig. 4.3).

Raising the pipe above ground on the VSMs did not ensure that there would be *no* thawing of the ground. The disturbance inflicted on the ground during the course of construction was sufficient to initiate temperature changes in the soil which could result in thawing to a significant depth, in all except the coldest, most northern areas. Climatic change too, might in places initiate a continuing thawing.

Ultimately, the solution of the problem lay in the so-called thermal VSM.[10] The thermal VSMs are equipped with devices known as heat pipes. These are sealed 2 inch diameter tubes within the VSMs. They extend below the surface and contain anhydrous ammonia refrigerant. In the winter months this evaporates from the *lower* end of the

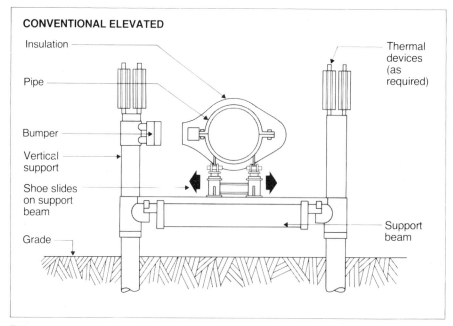

Fig. 4.3 Diagram showing vertical support members for elevated pipe. The VSMs extend into the permafrost so that they are not subject to frost heave in the active layer. The bumper is for possible earthquake induced movements. The 'thermal devices' are the heat exchangers of the heat pipes for preserving the permafrost.

tube and condenses at the top where there are metallic heat exchanger fins (Plates 9 & 10). The evaporation process occurs because, during the winter months, the ground is *warmer* than the air outside. The evaporation process itself cools the lower end of the pipe and the surrounding ground, and this is the point of the device: by cooling the permafrost in winter its temperature is sufficiently lowered to prevent thaw during the summer. The mean ground temperature falls, the heat pipes preventing the warming that would otherwise occur following disturbance of the ground surface. About 610 km of pipe were built above ground, and about 80 per cent of this length had thermal VSMs. The use of the heat pipes enabled much shorter 'legs' to be used on the VSMs. Typically the VSMs extended 8 m into the ground, although in some cases more than 20 m was necessary. In the absence of heat pipes the legs of the VSMs had to be overdesigned to allow for the deepest possible thaw that might occur.

The problems of the thermal contraction and expansion of the pipe itself, were overcome by a combination of the facility for movement at the VSMs, and the placing of the pipeline in a trapezoidal zig-zag (Pl. 9). The pipe was rigidly fixed at intervals of 250 and up to 600 m by anchor supports (Fig. 4.4). In general, the pipe can move laterally through some 4 m, and in special situations more than this in a vertical direction (Pl. 10). The allowance for movement also provides protection against the effects of earthquakes.

Early plans had called for wooden piles but it became apparent that VSMs would only react successfully to the various stresses and displacements if constructed of steel. In addition to the stresses caused by thermal contraction and expansion of the pipe, there were also those due to contraction of the permafrost during the winter, and evidenced by the cracking that leads to ice wedge polygons. In the end about 640 km of pipe, just about one-half of the line, were 'conventionally' buried and placed at depths of 45 cm to 4 m (Fig. 4.5). Buried along with the pipe are two zinc-ribbon anodes, to prevent the electrolytic corrosion otherwise to be expected with a buried steel object.

Plate 9 Alyeska pipeline in mid-Alaska. Use of the vehicle path, which is on the gravel surface mat, is very restricted to avoid collision with the pipe supports. The main service road which follows the line, is seen in the distance on the right of the picture.

Plate 10 The bend in the pipe is part of the provision for thermal contraction or expansion, and provision is made for movement of the pipe on the cross-support for the same reason. In this case, due to a miscalculation, lateral movement of the pipe has already brought it to the extreme position. If it cools it will move away, but if it warms further it will disturb the vertical support member. The light-coloured heat exchangers on the top of the VSMs are part of the heat pipe arrangement, serving to keep the ground frozen and thus anchoring the supports.

Some 7 miles of the pipeline were categorised as 'special burial' (Fig. 4.6). These were stretches where the above-ground situation was for one reason or another, especially to be avoided, but where 'conventional' burial would have resulted in serious thawing and subsidence. The 'special burial' sometimes involved merely insulation, but

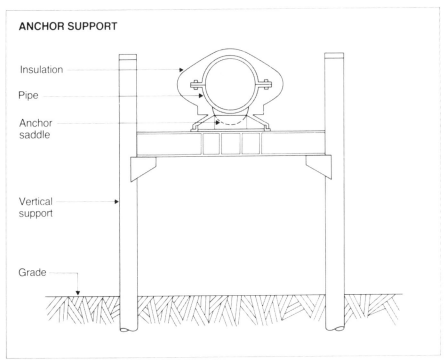

Fig. 4.4 Diagram of support members where pipe is fixed. There are normally four legs.

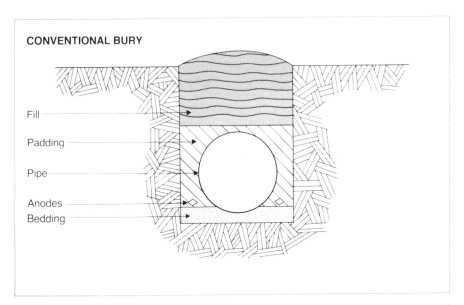

Fig. 4.5 Diagram showing pipe buried in ground. This mode is used where there is little danger of thawing of ground ice.

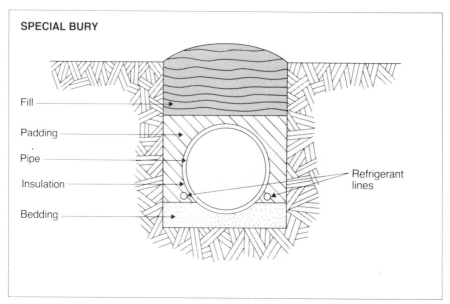

Fig. 4.6 External factors sometimes dictate that the pipe be buried, even though there is ground ice present. In these situations the pipe is insulated, and refrigerant is circulated to further counteract the warming of the ground. (Figs. 4.3, 4.4, 4.5, 4.6 are reproduced from Alyeska Pipeline Service Company 1977, with permission.)

often was an expensive arrangement in which the heat escaping from the pipe was removed by the cooling provided by refrigerant lines. The refrigerant, of course, required the installation of compressors, and 'special burial' presented a number of potential problems. If the amount of cooling is not correctly adjusted there is the possibility of thaw occurring, or, under certain circumstances, of additional freezing and frost heave. Since only a very small part of the pipeline was constructed in this fashion these difficulties are not expected to have far-reaching consequences. It seems reasonable to assume that a modified structure, whether of 'above ground' or 'buried' type, could be developed for these segments, even if at great cost. An interesting comparison may be made with the proposed Alcan gas pipeline. In that case burial is proposed for almost the entire length so the problem of techniques for controlling the temperature presented by the pipeline to the surrounding soil becomes much more critical.

Initially the Alyeska line was planned to pass under rivers, and it was in fact placed 5 to 6 m below the Tonsina river, after coating with 9 inches (22 cm) of concrete. Other large rivers, however, are crossed by bridges to which the pipe is slung. That over the Tanana River is 360 m in length.

In the case of river and stream crossings the problems were only partly due to permafrost. The scour of the rivers in spring means that the upper 2 or 3 m of the bed are periodically liable to be removed. If buried the pipe must be placed below these layers. The effects of lateral scour and erosion present even more of a hazard for, in time, they might expose the pipe or its foundations necessitating major remedial work. Many of the rivers traverse flat plains in which there are often several channels which constantly shift their course. Lateral migration of a few metres each year (by bank erosion and slumping) poses great difficulties for the design of the transition from the conventional

Plate 11 The Yukon river crossing in central Alaska. Here the pipeline (recognised by the zig-zag configuration, which allows for thermal expansion) and the accompanying road are carried on the same, specially constructed bridge.

burial, or elevated mode, to the sub-bed or bridge mode of construction. Permafrost is usually absent below larger rivers but can present special problems below the banks. The permafrost adjacent to the unfrozen region below the river will necessarily be relatively warm and prone to thaw. The burial of the pipe below rivers is of course favoured from an aesthetic point of view, but as design work progressed, the decision was made, for example, to traverse the frozen silts at the Gulhana River crossing with a 120-m tiered bridge instead. The engineering of the river crossings required not only avoidance of scour problems, which on occasion involved some damming or modification of flow patterns, but also the preservation of conditions suitable for the fish population.

Apart from the river crossings, a consideration of the effects on the natural drainage of terrain on either side of the line was essential. The problem of icings persists in that no standard or uniform procedure for preventing their build-up, or for removing the already accumulated ice, has been developed. Both the service road and the pipeline itself are the cause of certain icings. Techniques involving steam jets to thaw frozen culverts or to reduce an Aufeis accumulation, have proved costly and inefficient. Special drainage channels, or drains, aiding the flow of water during the autumn freeze-up have been constructed. When a growing mass of ice presses upon a VSM, an expensive and difficult removal of ice is essential.

Gas and oil storage units in the vicinity of Prudhoe Bay and elsewhere in the permafrost region, were elevated on piles 2 to 2.5 m above the ground surface. Smaller buildings were placed on gravel pads. The eight pumping stations[11] at present operating at intervals along the line are assemblages of buildings housing advanced gas turbines, which power centrifugal pumps, complex valve systems and storage tanks. In addition to the pumping function (which results in the pumping stations being located more closely on upslopes), the equipment can be used for pressure relief and temporary diversion of oil flow in the event of a leak. In several of the stations the foundations of the buildings have refrigeration systems to maintain the underlying ground in the frozen state.

The passage of men and equipment in the vicinity of the pipeline inevitably destroys vegetation, and initiates thaw of the permafrost. Large quantities of gravel were therefore placed on the ground surface as 'mats', and in particular a 1.5 m layer provided a foundation for the road which parallels the pipeline. Five to ten centimetres of foam plastic board was frequently placed under the gravel.

In spite of the extreme cold during the winter, construction in permafrost regions is normally carried out while the ground is frozen. During the summer, temperatures in central Alaska often exceed 25°C and even in the north of Alaska temperatures of 30°C can occur. Rapid thawing and a relatively deep active layer makes passage of construction vehicles difficult in wet or ice-rich areas. Damage to the ground surface is much greater than is the case during winter operations. The passage of vehicles in the winter usually involves construction of snow roads. These are prepared by compaction, but in northern Alaska the low snowfall meant that this was not always possible and, ironically, snow-making machinery had to be used.

4.6 THE COMPLETED PIPELINE

Oil started to flow in the Alyeska pipeline in July 1977 and has continued to do so with only minor interruptions. Viewed from the air, this silver thread on its dark supports, made conspicuous by the lighter background of bare gravels beneath, usually produces feelings of awe and excitement. Even those sceptical of the activities of the petroleum and pipeline companies, would be hard put to suggest how the intrusion on the natural surroundings could have been further reduced. Those privileged to see inside the compressor stations, or to examine the line closely from the ground are likely to feel that the concern for the 'environment' felt by some was misplaced. Construction has been achieved without the damage to terrain some envisaged. Today there is more likely to be surprise at the apparently undisturbed nature of the surroundings of such a large and recent structure.[12] Seeding and planting operations are further reducing the areas of exposed soils. The sometimes expensive provisions made for animals, especially caribou, to cross the line appear sufficient,[13] and indeed those most concerned with wildlife do not appear to have any great cause for concern.

The many uncertainties, doubts and objections that were voiced before construction was completed should not be forgotten. Even at quite late stages there were doubts as to whether much oil would actually arrive at Valdez. For the engineers of the pipeline there was an element of uncertainty and challenge, without which they might not have been driven to such achievements. There still remains the test of time. Only after several years will it be clear to what extent maintenance and modification are necessary. The line is continually monitored for changes due to permafrost thawing, erosion or other soil or terrain changes. Airborne infrared photography is used to constantly check the operation of each heat pipe by measurement of the temperature of its heat exchanger.

It has been argued that much of the escalation in costs resulted from the delays necessitated by the detailed examinations of quasi-legal and regulatory bodies. The

Plate 12 The Alyeska pipeline near Fairbanks, Alaska. In the foreground the pipe is buried to allow a highway (not seen) to cross it. In the distance it is again buried on the hillslope, where there is little danger of buried ice and resulting thaw settlement.

introduction of the Environmental Protection Acts in the United States is blamed for much of the delay. These arguments must be weighed against the fact that in 1969 plans were being proposed that were certainly unrealistic. Those that would have involved burial of the pipe for almost the entire length are an example. The subsequent thaw-settlement problems would have been irremedial. Of course, these and other potential dangers would have been recognised as the project developed, but changes might then have been costly, difficult, and damaging to the environment. It takes time to collate all the knowledge necessary, and to carry out the research demanded. It seems reasonable that this time was also used to examine issues which concerned other elements of society. However beneficial resource development may be in the long run, it has to be weighed against the negative aspects of such developments. Given the novelty of the construction, the technological and scientific uncertainties, it is hard to avoid the conclusion that the delays and examinations were essential to the successful construction of the Alyeska pipeline. The knowledge gained about the terrain and its interaction with the pipeline structures ensured good geotechnical designs and, at the same time, dispelled many of the doubts about the effect of the pipeline on its natural environment.

Even before construction of the Alyeska oil pipeline was commenced, a greater and even more challenging project was being planned: the Mackenzie Valley gas pipeline, to carry the gas being produced in the Prudhoe fields. The history of that pipeline is an even more interesting one from the point of view of the interplay of modern technology, scientific research, and the public interest.

Notes

1. A richly illustrated two-volume book (Allan 1977) on the Alyeska project was sponsored by Alyeska Pipeline Service Co. A more penetrating but easily readable account is that written by a business journalist (Roscow 1977).
2. Allen (1977). By comparison, drilling and exploration activities for the oil and gas itself, at Prudhoe Bay cost some $10 billion.
3. There have been repeated attempts at such analysis, and both the results and the implications drawn have been the subject of much controversy. The companies owning the pipeline (owner companies of the Trans-Alaskan Pipeline System) have themselves issued a detailed synopsis (TAPS 1977). Perhaps not surprisingly it ascribes most of the cost escalation to external economic factors, largely associated with enforced delays in construction.
4. Allen (1977).
5. The first calculations of thaw around a warm pipeline were probably by government scientists in the United States (Lachenbruch 1970), and Canada (Gold *et al*. 1972).
6. Roscow (1977).
7. A series of papers by Brown (e.g. 1964, 1968) of the National Research Council, Canada, describe for selected parts of Canada, the distribution of permafrost as it relates to particular vegetation, soil, and other terrain features.
8. Ice wedges are not always revealed by the appearance of the ground surface. Mackay (1970) describes one such situation in the Mackenzie Delta, where he estimates a square mile (2.55 km) may have 100 linear miles (160 km) of ice wedges. A road or pipeline will therefore cross wedges frequently, and if these go unrecognised in the design gross disturbance will ensue.
9. The term 'Arctic desert' is quite widely used, precipitation being regarded as the main criterion. The quite diverse terrain conditions associated with low precipitation are described in Smiley and Zumberge, 1974.
10. Excellent descriptive material, easily readable but fairly technical, is distributed quite freely by the Alyeska Pipeline Service Company. This is in addition to the sponsored book by Allen (1977), and attractive small but informative brochures which are given to visitors to the company's facilities. The Company's address is: 1835 South Bragaw Street, Anchorage, Alaska 99504 U.S.A.

11. These are sufficient for the current flow of oil. Four additional stations are planned for the expected doubling of the flow.
12. Alaskans will only have access to the region because of the road constructed by Alyeska; they will see a substantially less disfigured terrain than the hundreds of thousands of tourists who move through the northern European tundra regions. Electric transmission lines in similar terrain in Norway, for example, are far more conspicious and disturbing to the nature lover in an otherwise tidy country.
13. In the case of the Caribou herds it was feared that the appearance of the elevated line would deter them from passing under it, and they would thus be deflected from their migration paths. Apparently no such problems have arisen; the special burial procedures with refrigeration were utilised at several sites to provide a clear path for the herds.

CHAPTER 5

THE GAS PIPELINES AND THE FROST HEAVE PROBLEM

The process of tapping oil brings natural gas to the surface. In the past, this gas was commonly flared off. In North America at least, such waste of energy resources is now viewed adversely, and the gas is normally pumped back into the ground where it also serves to maintain pressure in the well. In the case of the Prudhoe Bay field, it was planned that natural gas would be produced for sale after four or five years operation of the Alyeska oil line. Examination of methods for bringing the natural gas to markets in the United States thus got under way in the early 1970s. These examinations were spurred on by further gas finds, particularly in the Mackenzie Delta, and also by the apparent risk of gas shortages in the United States a few years hence.

5.1 THE MACKENZIE VALLEY PIPELINE

In March 1974, Canadian Arctic Gas made formal application to the Government of Canada, for the building of a 48 inch diameter gas pipeline down the Mackenzie Valley, and into the United States. At its northern end, it would be connected to Prudhoe Bay, by a link across the northern Yukon (Fig. 5.1). Plans for transporting the oil by the Alyeska line to the south Alaska coast were already well advanced. The gas pipeline was also intended to carry gas from the fields in the Mackenzie Delta – fields, incidentally, which were not fully proven, and which have, to date, proved to have smaller reserves than then supposed. Later in 1974, Foothills Pipelines Co. made application to build an alternative line, of 42 inch diameter, down the Mackenzie Valley, but to carry gas only from the Mackenzie Delta. The Canadian Government established the Mackenzie Valley Pipeline Inquiry to examine the proposals; subsequently this became a judicial inquiry, under Mr Justice Thomas R. Berger.

Whether a pipeline is to carry gas, or oil, is fundamental to its design, and thus, these proposals for northern gas pipelines meant new technological demands. While an oil-carrying pipeline is generally warmer than the soil adjacent to it because of the fairly high temperatures of the oil from deep wells, gas has a low volumetric heat capacity and quickly drops in temperature. If compressed, however, its temperature rises, and if it expands, it falls (the principle of the domestic refrigerator). Because the rate of transport is increased if the gas is denser, gas is compressed by compressor stations at intervals along a pipeline. In travelling along the pipe between the stations, however, some expansion, and consequently cooling, occurs.

Both Mackenzie Valley proposals involved further, deliberate cooling procedures for the gas such that the pipe itself would be at temperatures below freezing for a

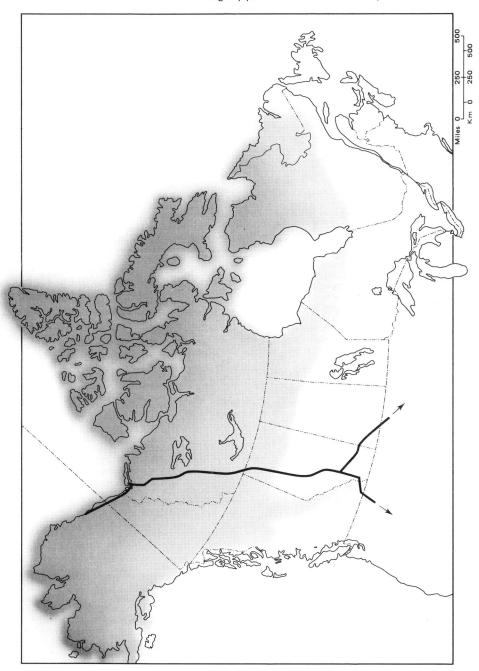

Fig. 5.1 Location of proposed Mackenzie Valley Gas Pipeline.

significant part of its length. The chilled pipe could be buried in permafrost, without danger of thaw settlements. A buried pipe appeared to offer substantial advantages of security, and ultimately, minimum disturbance to its surroundings. It appeared that the chilled buried pipeline was not only just what was required technologically, but also the most attractive solution from the environmental protection aspect. Burying the line was better than laying it on the ground surface, or elevating it on piles as had been done with much of the Alyeska oil line.

Most pipelines have the gas under rather high pressures. A remarkable feature of the Mackenzie pipelines was that the proposed pressures were even higher than usual. The maximum operating pressure according to the Arctic Gas proposal was to be 1 680 pounds per square inch (that is, 11.58 megaPascals). The gas then presses against the pipe with a force similar to that which the sea exerts on a bathyscaphe at about 1 100 m depth. The pressure is 55 times that in an average motor car tyre. The essential design requirement is that the possibility of rupture be held to a minimum. Clearly, too, the higher the pressures the more stringent are the design conditions.

A large diameter buried pipeline containing 'chilled' gas was an innovation that, apparently, had not been the subject of any substantial consideration even by Russian specialists. The low temperatures of the pipe would cause additional freezing of the soil surrounding the pipe. This leads to the 'frost heave problem' as it relates to northern pipelines.[1] The problem is not so easily defined, however, as in the case of frost in highways, and certainly not so easily managed.

Indeed, the nature of the problem and the proposals for solving it make a curious story; of scientific evidence that went unnoticed, of construction designs confidently prepared, and of a chain of circumstances culminating in an abrupt change of plans.

5.2 FROST HEAVE AND THE COLD PIPELINE

There were particular situations along the pipeline route which posed especially critical problems, for example, river crossings, or unstable slopes, but frost heave was a more general problem. Not only were greater lengths of the line likely to be exposed to this hazard but frost heave is also involved indirectly, in many of these other terrain problems. Indeed, in order to understand northern terrain an understanding of the frost heave phenomenon is essential. The main question, however, was whether freezing soil as it heaved could bend the pipe, leading to the risk of rupture.

In regions of discontinuous permafrost, the chilled pipe would necessarily cause freezing of the soil as it traversed the pockets of unfrozen soil. As years passed, a thicker and thicker ring of frozen soil would form around the pipe. The designers realised that this freezing would frequently result in heaving. To counter this the consultants to Arctic Gas proposed weighting the pipe down with an earth mound – a berm (Fig. 5.2). Such a berm would, they said, only be necessary for certain stretches, and would never need to exceed about 3 m in height.[2] This height represented the maximum significant 'lifting' ability or heaving pressure that freezing of the ground was thought to generate.

Because of its importance in highway and other foundation design, heaving pressure has been the subject of many research studies during the last 40 years. Many rather simple field investigations have been made by highway authorities, but the problems often require a more sophisticated and detailed approach. An unexpected aspect of the problem which relates to permafrost areas is that soil already frozen may, on being further cooled, undergo additional heaving. Such heaving takes place very slowly. This is one reason why laboratory experiments with freezing soils are extremely

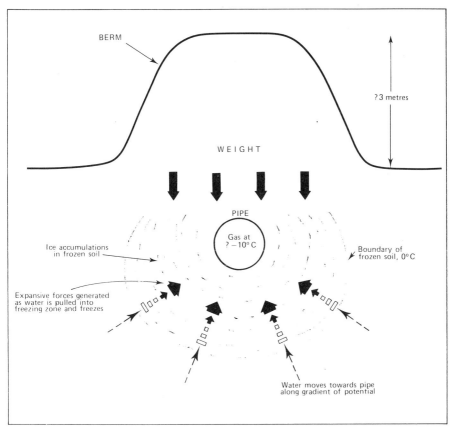

BERM

?3 metres

WEIGHT

PIPE

Gas at
? −10°C

Ice accumulations
in frozen soil

Boundary of
frozen soil, 0°C

Expansive forces generated
as water is pulled into
freezing zone and freezes

Water moves towards pipe
along gradient of potential

Fig. 5.2 Diagram of effect of cold gas pipeline on surrounding soil. As water tends to move towards the pipe, because of the temperature gradient, the soil tends to heave. According to the 'shut-off' theory, the weight of a sufficiently large berm would overcome the heaving forces, and the pipe would remain in place.

difficult. The need for very precise control of temperature and for mechanical devices to measure expansion, pressure, and the movement of small quantities of water into samples, are yet further difficulties for the experimenter. The results may be equivocal, or it may not be possible to devise an experiment to answer exactly the question asked.

These difficulties have been evident in the study of the heaving pressure. Although earlier experimental work had included measurements of heaving pressures, it was not easy to predict the pressures to which the pipeline might be subjected by heaving soil. The forces with which water is pulled into a freezing soil sample could, it seemed, be quite large. Logic suggests that the resulting expansive forces, or pressures, would be as great.[3] But then, nobody had actually shown such pressures were maintained while a significant amount of heave occurred. It was known that heave was always slow when the pressures were high, but as freezing around the pipeline would continue for 30–50 years there would be plenty of time. By the early 1970s little was known about the magnitude of the highest pressures generated by frost heave.

In such a situation a combination of theory and experiment is sometimes the best answer, but for the major pipeline proposals these answers appeared late in the day and did not make the constructor's task any easier.

5.3 ORIGIN OF THE HEAVING PRESSURE

By the late 1960s the view shared by most scientists and engineers who studied soil freezing, was that as the frost penetrated the ground, it was the size of the soil particles and of the soil pores which determined the 'water-attracting' or 'suction' forces. This limited understanding was the result of years of scattered experiment and practical experience – the pace of applied science can be sluggish. Only a few had seen the connections[4] with the work of Laplace, Kelvin and Clausius-Clapeyron, which was outlined in Chapter 3. When ice forms in very small spaces such as soil pores, capillarity (as Laplace's work indicated) and adsorption forces (from the mineral particle surfaces) result in the ice maintaining a higher pressure than the adjacent water. The difference between the two pressures is greater, the smaller the pore, or, as the space for water is reduced. At the same time the freezing temperature is lower than normal. The ice presses on its immediate surroundings, and the pressure of the ice is, in fact, the heaving pressure exerted by the soil as a whole. That is, the pressure of the ice corresponds to the weight effect, or resistance of the overlying material.

The key fact, high pressure in the ice, low pressure in the water, is of course an 'unusual' one, and has its origin in sub-microscopic, molecular effects themselves brought into play by the special environment *within* the soil pores. Ice in lakes, or in drinks, does not behave in this way. The low pressure in the water constitutes the 'suction' pulling water into the pores from outside. The difference in pressure $P_i - P_w$, increases as the temperature falls (Fig. 3.4), so that the 'suction' is *greater*. If the weight (from the overlying material) acting on the ice and constituting P_i, is increased, on the other hand, then P_w will be higher too by the same amount. That is, there will be *less* suction effect so that less, or even no water is drawn into the soil, and the heave is reduced accordingly.

As these processes were gradually understood in the 1960s, however, a curious misconception persisted, even amongst those scientists with the greatest experience. A few made reference to the possibilities for movement of water within the frozen ground,[5] but it was usually assumed that the water attracted by the freezing ground accumulated as ice right at the boundary of the freezing soil. Highway engineers in particular, visualised the frost heave occurring right at the lower edge of the frozen layer, that is, at the boundary of the frost as it moves down in the ground. At that point, of course, the temperature is near to 0°C.

Perhaps the misconception was not surprising. That water can move *within* the frozen ground, and ultimately freeze at some point, even perhaps far in the already frozen layer, seemed unlikely, if not a contradiction in terms. Secondly, frozen soil, like deep-frozen meat, is such a rock-like material that the idea of it expanding, pushed apart from the inside by accumulating ice, seems far-fetched.

Since Nersessova's work however (see Ch. 3), the existence of significant amounts of water alongside ice in the frozen soil had been quite widely described. The explanation had been provided too: as a soil continues to cool, ice forms from water left in progressively smaller and smaller pores and crevices and layers around particles. Thus the correct picture of a frozen soil is of mineral soil grains surrounded by a network of ice, but without the grains actually being encased in ice. Rather there are, between the ice and grains, interconnected films and pockets of water. It could be expected that the water in these films is able to move slowly whenever there is some force tending to make it do so. Because the freezing is going on in smaller and smaller spaces, the attractive forces, the suction, in the water become greater and greater at lower temperatures. So does the possible heaving pressure. It is the difference $P_i - P_w$ which gets bigger as the temperature gets lower (Fig. 3.4). Of course this is what Clapeyron's and Kelvin's work suggests, and it might well have been foreseen in advance of the experiments of Nersessova and others.

Thus we have a picture, a model, of water *within* the frozen soil, tending, as the temperature drops, increasingly to attract more water to it. In scientific terms we say that the water has a 'potential', more precisely, a lower potential than water elsewhere. The concept of potential is closely related to that of 'free energy', the concept introduced by Willard Gibbs (Ch. 3). It is differences, gradients, in potential that make water move. It moves from positions of higher to lower potential. Other examples of differences in potential are hydrostatic pressure differences, elevation differences (water flowing in a river), osmotic pressure effects (water moving from ground, up a tree). Figure 3.4 implies that, relative to 'ordinary' water, the potential of unfrozen water, for example, in a frozen soil (without any load on it) at −1°C, is *lower* by an amount equivalent to some 12 times atmospheric pressure or 1 200 kPa.[6] At −2°C the amount is 24 times. Thus the forces tending to drive ordinary water (at atmospheric pressure) into soil *to the point* where the temperature is −1°C, are as great as those that would push water 120 m up a vertical pipe. If more water (for example, at atmospheric pressure), was able to reach *this* frozen soil layer, then the ice forming in the soil would expand to lift even a 60 m column of soil (soil is approximately twice as heavy as water). The investigation of this possibility came to have an unusual significance.

5.4 FROST HEAVE AND THE SHUT-OFF PRESSURE

Canadian Arctic Gas supported their application to build the Mackenzie Valley line, with the results of a substantial programme of research. To investigate frost heave the consortium of consultants[7] established field trials using 100-foot lengths of full scale, cooled, and buried pipe, and a laboratory where specially modified soil-testing equipment enabled numerous 4 inch (10.16 cm) diameter soil samples to be observed under controlled freezing conditions. As geotechnical research, these investigations were probably more comprehensive than any ever undertaken, in Canada, specifically for a single engineering project. The researchers set out to investigate the magnitude of the frost heave pressures and demonstrated that, even at their maximum, these pressures could be overcome by the use of berms. The berms would be necessary only at a limited number of sites, because it would be possible to reroute the pipe around the worst places.

If the maximum heaving pressure were 200 kPa, an equal counter pressure would be provided by 10 m of soil. The experiments indicated that this would probably be the worst that need be expected considering all the types of soil and conditions that might occur. The consultants' reports (which were made public through the Berger Inquiry) refer to 'shut-off pressure' rather than 'maximum heaving pressure' although the terms are essentially synonymous. In the laboratory experiments, enough pressure was applied to the sample to just stop its further expansion, or heave, thus the pressure applied equalled the maximum expansive pressure of the sample. It is the absorption of water into the sample (giving heave) which is 'shut-off', when the expansion of the sample is prevented.

The idea of measuring 'shut-off' pressure was not in conflict with findings of earlier investigators, nor was it entirely original. The assertion that the investigations established the greatest heaving pressures, of such a magnitude as to be just manageable, seemed questionable particularly as earlier findings suggest otherwise.

5.5 MEASURING THE MOVEMENT OF WATER THROUGH FROZEN GROUND

Although indirectly related, there was sufficient evidence in the scientific literature to suggest that the heaving pressures would be several times those predicted by the consultants on the basis of their investigations. On this basis earth constructions of 20 m or even higher might be needed to weigh the pipe down.

The possibility of reconciling these earlier scientific reports with the findings of the consultants' investigations was considered. An important factor was the slowness with which water might be expected to move within frozen soil. If the water movement and heave was restricted to the margin of the freezing soil, then the pressures would not be very great. It might take years for even a small quantity of water to accumulate within the frozen soil around the pipe, by passage through the tiny channels of unfrozen water adjacent to particle surfaces. If this were the case, the heave, within the lifetime of the pipeline, could be so small as to be of no significance. Even a 48 inch diameter pipe can tolerate some displacement, and the pressure becomes irrelevant if the heave is small enough. The analogy can then be made to a mighty fist, and an extremely slow motion punch – which however, is not carried through.

Given the presence of the films of water in frozen soil, and the likelihood that this water can move, how easily can it do so? Measurements of permeability, or hydraulic conductivity as they are also called, are made very often by ground-water geologists, geotechnical engineers and others. Quite simple devices are used in which, essentially, water lying on top of a small piece of soil passes through it to a measuring container beneath. Of course if the sample is frozen the experiment is, it would appear, unlikely to succeed because the water would either thaw the soil, or itself freeze. Amidst the freezing or thawing it would be extremely difficult to detect small amounts of water passing *through* the frozen soil.

While the massive pipeline research programme was underway, quite independently Timothy Burt, a young university researcher, was tackling this question of water movement in frozen soils as research for a Master's degree thesis.[8,9] He developed a simple apparatus in which one could actually see the contents of one container or 'reservoir' being augmented at the expense of another (Fig. 5.3), while the sample lying between them was frozen 'solid'. A quantity of sugar had been dissolved in the water in the reservoirs, this prevented the water in the reservoirs from freezing even though a degree or so below 0°C, the temperature of the frozen soil.[10] To initiate flow the pressure on one reservoir was raised. The sugar did not travel far into the soil – the water moves far too slowly for that, and a fine-pored membrane (Fig. 5.3) further hinders the sugar molecules. But it was equally clear that water was edging, very slowly, through the sample, since an easily observable quantity (measured by movement in the narrow tube, Fig. 5.3) was coming out of the 'far' side. When the pressure applied to the first reservoir was, for example, 30 kPa (the same as that given by a 3 m head of water) water might travel in the tube leading from the second reservoir at 0.5 cm per minute. The rate varied from soil to soil. For some soils at temperatures just below 0°C, the permeability coefficient,[11] is similar to that for certain dense and little porous materials when unfrozen (Fig. 5.4).

Now consider the gas pipeline, initially many degrees colder than the ground in which it is buried. After a while the temperature of the ground around the pipe is reduced. The ground may be already frozen, or it may form an annulus of frozen soil. The potential or 'suction' of the water resulting from the temperature gradient, will tend to move the water *towards* the pipe (cf. Figs. 3.4 and 5.2). The driving forces are easily

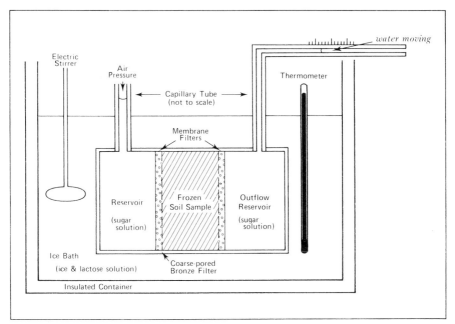

Fig. 5.3 Diagram of apparatus used to measure flow of water through a frozen soil. The water is caused to flow by raising the pressure of the water in the left-hand reservoir.

evaluated (using the Clausius-Clapeyron equation, and the various experimental findings) and are comparable to the effect of the applied pressure in the experiment (except that the forces can be very much larger).

From the results of the experiment it can be calculated that, in 30 years or so, so much water would have moved through the frozen ground towards the pipe and accumulated under it as ice, that the pipe would be lifted a metre or more at some places. The weight of a 6 m berm would not hold it down, even the weight of a 60 m berm might be insufficient. Furthermore the lifting would not be uniform over long disturbances, at some places it might not be lifted at all: the resulting bending of the pipe would be unacceptable. In fact the pipe would probably break.

The results of one experiment or set of experiments should not be relied on too heavily to prove or disprove a contention of such importance. It should be remembered that permeabilities of natural soils, when not frozen, range over several orders of magnitude (sand has a value of about 10^{-3} m s^{-1}, some clays 10^{-10} m s^{-1}). Even 'good' measurements are sometimes not accurate to an order of magnitude. There was no obvious way to check the results of Burt's measurements, and the uncertainty was critical.

5.6 A DIFFERENCE OF OPINION

The staff of the Mackenzie Valley Pipeline Inquiry recognised the possible problem of frost heave and initiated thorough examination of the designs and supporting documents. Counsel for the Government of Canada subsequently introduced the matter

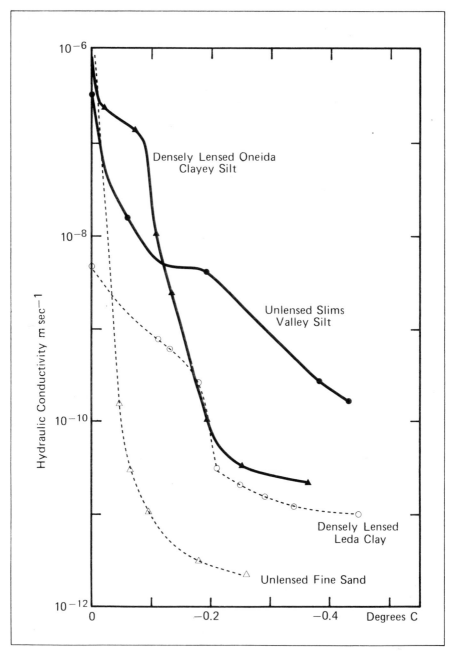

Fig. 5.4 Values of the permeability or hydraulic conductivity coefficient for various soils, over a range of temperatures. (From Burt and Williams, 1976.)

into hearings before Mr Justice Berger. The author was called as a witness in this connection, and put forward the opinion that there was a serious risk of disruptive frost heaving of the pipe.[12]

It is, of course, relatively easy to cast doubt upon the validity of experiments and even upon the integrity of engineering designs. It is much harder to prove that there is no significant risk of difficulties arising during the 30-year, or longer, life of a novel geotechnical construction. At any rate the suggestion of disruptive frost heave was vigorously opposed. Counsel for Arctic Gas called a panel of expert witnesses, all directly involved in work on the pipeline project. This panel provided a rebuttal,[13] which together with the original testimony occupies some 200 pages of the transcript. The belief was expressed that it was the suction forces that occurred at the frost line that determine the heave, not those within the soil. Furthermore, it was maintained that the suction forces pulling water into freezing soil had never been measured as exceeding one atmosphere (100 kPa).

There are, in fact, a few cases reported of water being pulled into the frozen soil in spite of 'shut-off' pressure applied to the soil equivalent to the weight of 20 m of soil. This of course implies suction forces of similar magnitude. But little water actually moved into the soil in these cases.[3] The question of the maximum suction remained uncertain. It was further argued that the presence of ice lenses would essentially block the movement of the water in already frozen soil. While it was agreed that the determinations of permeability made by Burt were not unrealistic so far as they went, it was thought that the values were so small that little migration of water would actually occur in the frozen ground in the field conditions.

That a layer of ice, extending right across a sample, would prevent the movement of water through the sample, seems reasonable enough. A large number of discrete ice layers (rather than a single one) could be expected to have a similar effect in a larger body of soil. Thus it seemed that movement of water within the frozen ground towards the pipe would, indeed, be slowed or stopped as soon as ice lenses formed. This point had, in fact, already been considered in rather interesting circumstances.

In the course of Burt's work, he had found that the flow of water was *not* blocked even by a disc of ice covering the entire cross-section of the frozen sample. This had caused concern because it suggested the experiment was faulty. It was feared that the flows of water, that had been observed, were not in fact passing *through* the frozen sample. How could they be, if the flow continued in spite of a blockage right across the sample? Perhaps there was a technical deficiency, a leak perhaps, around the sides of the sample.

Resolution of the problem lay in an experiment carried out eight years earlier by R. D. Miller of Cornell University (actually while on a sabbatical leave at the Norwegian Geotechnical Institute).[14] Miller had shown that, in an arrangement not unlike that in Fig. 5.3, when pressure was applied to water on one side of a disc of ice, a slow flow of water occurred away from the other side. The experiment was referred to as the 'ice sandwich experiment' and was seen as the possible basis for a water desalinisation process. If this process occurred with the ice lens in the soil, then indeed it would have the same effect as if the water went right through the ice.

Miller and his co-workers have developed these ideas and proposed that in frozen soils not only the water but also the ice moves slowly through the pores.[15] The process is one of accretion of molecules on one side of the ice within the pores, and the loss of molecules (by their transfer into water) on the opposite side. Thus any point within the soil ice is being slowly edged forward. Clearly, as this process involves freezing on the one side and thawing on the other, scientific description must involve both heat flows and temperature differences, and moisture flows and potential differences. If this process occurs, then it might be expected that, even where there are numerous layers of ice in the ground, a gradual movement of water (and ice) will result in an accumulation in the colder parts of the ground.

Presented with two conflicting standpoints Mr Justice Berger ruled that the parties should attempt to resolve their differences by private, written exchanges during the winter 1975–76. The two points of view can be stated briefly as follows: either water (and maybe ice) moves *within* frozen ground towards regions of lower temperature, where it accumulates, to give great expansive pressures or, water may migrate only to the edge of frozen soil, and accumulate there, which would involve only small expansive pressures. If the pressures were small they could be resisted and the heave would be prevented. Inevitably, the question of frost heave pressures posed a threat to the acceptance of the designs involving berms to hold the pipe in place. The lawyers for Arctic Gas, whose task it was to present the case for the acceptability of the designs to Mr Berger's Inquiry, had therefore to try to show that the suggestions of high heaving pressures were ill founded. Unfortunately, this desire coloured the written exchanges which were formal, even antagonistic, and this detracted from their value.

5.7 A CHANGE OF PLANS

In spite of the vigorous defence put up by Arctic Gas for their point of view, the questions were still being considered within the consortium of consultants for the pipeline. Several scientists with experience in the study of soil freezing were working for the companies involved. They were aware of the fledgling state of knowledge about long-term frost heaving, they themselves had contributed significantly to this knowledge.

The first public intimation that all was not entirely clear was an unexpected announcement, in the closing sessions of the Berger Inquiry by counsel for Arctic Gas[16]:

'I want to speak to you about the question of frost heave . . . We have recently discovered a malfunction in the apparatus used in connection with tests to determine frost heave effects . . . along the pipeline route.'

Counsel continued at length, and concluded by saying:

'Arctic Gas will take the position that in the matter of frost heave this inquiry is not in any position to make any specific findings in this regard and the issue will have to be settled to the satisfaction of the National Energy Board.'

Mr Justice Berger: 'Well it would be nice to know what the malfunction was.'

Counsel: 'What it was?'

Mr Justice Berger: 'Was it the equipment used to test? Is that the problem? It's not a theoretical difficulty, I gather?'

Counsel: 'Oh no, no. Not at all. Not at all. . . .'

The significance of the statement is open to misinterpretation. The malfunction was, in fact, a leak which occurred in a number of the chambers or frost heave cells, in which samples were being tested at the consultants' laboratories. Air under pressure seeped into part of the volume measuring system, such that the 'shut-off pressures' observed (the maximum heaving pressures the freezing soil would exert on the pipe) were substantially lower than was really the case. The leak was discovered because one of the test cells had been lent to the Division of Building Research, National Research Council of Canada, where investigations along related lines had been carried out for many years.

It is not unusual to find flaws in experimental apparatus, which invalidate a particular set of experimental results. Research is often a process of two steps forward and one step back. The laboratory and field tests for the Arctic Gas project were a significant contribution to knowledge of soil freezing and frost heaving, and were backed by resources normally only available for research with direct applications.[17] Difficulties with an experiment, and the fact that certain reported findings had to be modified, serve as a reminder of the fallibility of all scientific research, of the inherent tendency to delays and setbacks, and of the need for assessing results in the context of other studies. The situation can be compared with that of Poynting's mistake, referred to in note 13, Chapter 3. Had no leak been discovered moreover, the findings in the laboratory experiments should not have been accepted by themselves as sufficient proof that there would be no major frost heave problem.

It had nevertheless been stated by counsel for Arctic Gas that there was no 'theoretical' difficulty. The implication was, presumably, that the heaving pressures would be sufficiently low as not to disrupt the pipe, when buried in the ground under (where necessary) the weight of an earth mound. Equally though, it could be taken to mean that the theoretical concepts, concerning soil freezing, held by the consultants had not changed. Here perhaps was a distinction that should have been made earlier between the 'corporate' view and that of individuals.

In late October 1976, Mr Berger wound up the hearings of his commission and turned to the preparation of his two-volume report. The first volume appeared in May 1977, to become, with its lucid text and evocative illustrations, a best-seller in its genre. The report makes it clear that the frost heave problem was not solved, and that consequently it was not possible for the inquiry to take a standpoint with regard to the acceptability of the proposed designs for the construction. The designs were not complete and changes would raise new problems. It seemed premature to Mr Berger that the Canadian Government should give unqualified approval for a right-of-way, or provide financial guarantees.[18]

In the meantime the process of examining the proposals of the two pipeline consortia had continued, but in the less colourful, more bureaucratic atmosphere of the National Energy Board of Canada. The function of the N.E.B., a regulatory agency, is to examine proposals and plans, and ultimately to authorise the issuance of the certificate permitting construction of pipelines. The particular social, and environmental questions relating to the Mackenzie Valley proposals were novel. The Board's inquiries had generally been of a technical, legal or economic nature. The N.E.B. hearings were not such as to encourage the participation of concerned citizens. There was no steady stream of witnesses, 'expert' or otherwise, from universities, citizens' groups, or affected communities; but the presentations of the competing pipeline consortia were thoroughly examined.[19]

The Board decided against the construction of a Mackenzie Valley pipeline. In introducing the reasons for its decision, it reported:[20]

'Never before has the Board been faced with such a complex and difficult task in making a decision on applications before it. This is not only because of the immensity of the projects themselves and their importance to all Canadians, but also because of the magnitude of the potential socio-economic impact on the peoples of the north and the critical concerns related to the protection of the Arctic environment. Confounding the situation were late filings, introduction of the

Foothills (Yukon) application part way through the proceedings, amendments of applications, the eleventh-hour filing of applications for a 48-inch Foothills (Yukon) alternative [to the Mackenzie] system, and research still in progress on the question of mitigative measures to offset frost heave and thaw settlement in the sensitive discontinuous permafrost areas . . .

The reference to research still in progress relates to the question of maximum heaving pressures. Subsequent to the discovery of the leak in the test apparatus and the assertions about the absence of theoretical difficulties, further tests had been reported to the N.E.B. by Arctic Gas. These tests 'showed that shut-off pressures may be greater than 7 000 to 10 000 p.s.f.' (334 kPa to 479 kPa). The test data, in fact, revealed that 'shut-off' was not obtained in some cases at more than 10 000 p.s.f. This implied that to weight the pipe down would require a 30 m (100 ft) berm, 'or burial at a depth of about 140 ft, or some corresponding equivalent combination of berm height and burial depth'.[21] Such construction procedures were clearly not a practical possibility.

Thus at a late stage in the hearings, in February 1977, Arctic Gas were forced to present proposals for redesign, in fact, a series of design solutions, to be applied to perhaps 400 km of the Mackenzie Valley pipeline passing through the discontinuous permafrost. Obviously the effectiveness of the new designs was unproven, and discussion of them is now more relevant to the Alcan pipeline, to which, notwithstanding its reservations about permafrost and frost heave, the National Energy Board gave tentative approval.

NOTES

1. The effects of continuous freezing temperatures applied to the ground are the source of the 'permafrost' problems around liquid natural gas storage tanks at Canvey Island in the U.K. The problem there is due as much to thermal contraction at these very low temperatures as to frost heave.
2. National Energy Board (1977) Vol. 2, Chap. 3, p. 60. The load effective on the pipe was increased by the pipe being buried below original ground level (Fig. 5.2).
3. Using simple devices to measure the pressure of the water being drawn into the soil and to load the soil, Beskow (1935) demonstrated that heaving pressures could be 100 kPa, that is, approximately equal to the weight effect of 5 m of soil. By a process involving an initial raising of the surrounding air pressure, Williams (1967) demonstrated suction forces of more than 400 kPa. The procedure, however, involved measuring the force required to just stop the flow of water to the sample.
4. For example Edlefsen and Anderson (1943) – a classic theoretical work in soil thermodynamics, which nevertheless was partly erroneous with respect to soil freezing.
5. In particular, detailed theoretical studies by Harlan (1973) led to the hypothesis (Harlan 1974) of slow water movements in permafrost due to temperature gradients.
6. The pressure of the atmosphere is no longer the basis of an accepted unit of pressure. The 'standard atmosphere' is approximatelly 100 kilopascals (10^5 Pa, or 100 kPa, see note 5, Chapter 3).
7. Consultants to Arctic Gas in the design of the pipeline were Northern Engineering Services, a consortium of several prominent western Canadian foundation engineering consultant firms.
8. Burt was working under the author at Carleton University in Ottawa. He had come from Cambridge University, seeking added excitement by pursuing further studies abroad.
9. Burt (1974), Burt and Williams (1976). Other relevant works were by Hoekstra (1966) who showed increase of moisture content along a temperature gradient within frozen ground, and Harlan (1973) whose theoretical studies clarified the quantities involved.
10. In more scientific terms, the concentration of the solution was adjusted to have the same free energy as the unfrozen water in the frozen soil sample. The solution does not freeze, and nor does it tend to enter the sample – until the pressure is raised on one reservoir.

11. The permeability coefficient, widely used in soils engineering, represents the rate at which water would move through a soil, if there were 1 m head of water for each m of soil travelled (see e.g. Craig, 1976).
12. Mackenzie Valley Pipeline Inquiry (1975), Proceedings, Vol. 69, pp. 10 234–10 375.
13. Mackenzie Valley Pipeline Inquiry (1975), Proceedings, Vol. 72, pp. 10 820–10 947.
14. Miller (1970), R. D. Miller, a professor of Soil Physics, is, by background, a soil scientist concerned with agricultural questions. An occasional interest, many years ago, in soil freezing, had been followed by a series of important studies of the topic carried out in conjunction with a small group of advanced students.
15. Miller, Loch and Bresler, 1975; Miller, 1978.
16. Mackenzie Valley Pipeline Inquiry, 1976, Proceedings, Vol. 195, pp. 30584–30585.
17. '. . . the detailed experience gained by those associated with the Mackenzie Valley Pipeline is unusually valuable both in itself and as a base for further necessary investigation and studies. . . .' Berger, 1977, Vol. 2, p. 134.
18. Berger (1977), Vol. 1, pp. 19–22.
19. There were 110 intervenors or interested persons involved in the hearings. At least seventy were companies, associations, etc., favouring early construction of a pipeline. Some eighteen intervenors were opposed to a pipeline at this time.
20. National Energy Board, 1977, Vol. 1, Chap. 1, p. 57.
21. National Energy Board, 1977, Vol. 2, Chap. 3, p. 63, 'p.s.f.' is pounds per square foot, a unit of pressure becoming obsolete (1 p.s.f. = 47.88 Pa).

CHAPTER 6

THE ALCAN PIPELINE

Mr Justice Berger's report on the Mackenzie Valley Pipeline proposals, recommended that approval should be withheld for at least 10 years. This would give time for full consideration of the implications of such a pipeline for native rights, and for the local economic and social conditions in general. It would allow time for measures to reduce possible harmful effects of the pipeline. The report made it quite clear that uncertainties remained over questions relating to the geotechnical design, especially frost heave. Although these were not the main reasons for the recommended delay, it was assumed that during the 10 years further research and an increase in understanding of the behaviour of freezing and thawing soils would resolve the uncertainties. In turn, this would lead to progressive modification of geotechnical designs. With the publicity and debate surrounding the Berger Inquiry, few people in Canada realised that there was an alternative northern gas pipeline proposal in the wings, which would be pushed forward with much greater urgency.

6.1 THE APPROVED PIPELINE

Mr Berger's report had favoured a proposal for bringing Prudhoe Bay gas to California by means of a pipeline that would roughly follow the route of the Alaska highway (Fig. 6.1), although the Mackenzie Valley Pipeline Inquiry did not examine the proposal in detail. The proposal, which had already been judged, by the Federal Energy Regulatory Commission[1] in the United States, to be as acceptable as the Mackenzie Valley line, involves paralleling the Alyeska oil line southwards to near Fairbanks and then proceeding through the southern Yukon. The route avoids the northern Yukon coastal plain, which Berger described as a unique wildlife habitat where pipeline construction should not be allowed under any circumstances. It also avoids, for the most part, lands which are the subject of dispute regarding native rights. It passes through regions having some risk of earthquakes, in the Yukon, and, to a lesser extent, in Alaska.

An alternative to an overland pipeline had also been considered by the F.E.R.C. This was the El Paso tanker route, which would duplicate, for gas, what has been achieved for oil – a pipeline across Alaska and shipment by tanker to the west coast. The recommendation of the F.E.R.C. was that a totally overland route be selected, subject to appropriate agreement by the Canadian government.[2] In addition to hearing submissions concerning the social, environmental and economic implications of the alternatives, the F.E.R.C. heard from many groups with a vested interest in one or other alternative.[3] The largest such 'group' might, of course, have been Canada itself, which

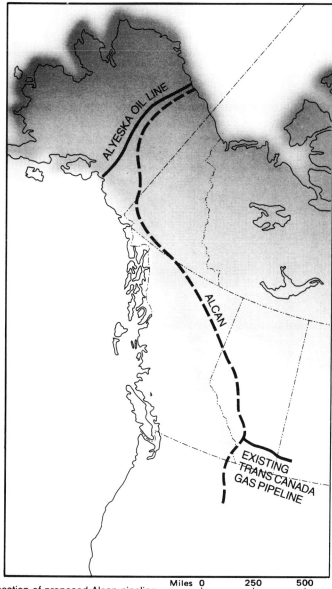

Fig. 6.1 Location of proposed Alcan pipeline.

Miles 0   250   500

Km 0   250   500

would be essentially excluded from any benefits of the 'El Paso' proposal. By the time the Berger Report appeared, it was clear that the United States government intended to proceed as early as possible with one or other alternative.

Only a few months after the Berger report's appearance, the National Energy Board of Canada made clear that there would be no Mackenzie Valley pipeline in the foreseeable future.[4] Instead, the N.E.B. supported the 'Alcan' proposals (as the Alaska highway route became known). Shortly afterwards Parliament in Ottawa gave its approval in principal to that pipeline, thus ensuring Canadian involvement in the massive undertaking to bring northern gas to southern markets. A new consortium of pipeline companies (essentially the 'Foothills' group in Canada[5] and 'North-West' in Alaska) thus came to have approval in 1977 to proceed with the construction of a 48-inch gas pipeline through the most demanding northern terrain. Subsequently this was changed to 56-inch diameter. The time anticipated for completion of the undertaking was scarcely longer than that envisaged for the original, ill-fated Mackenzie Valley line.

At the moment, there is only one factor which is likely to delay the commencement of construction of the Alcan line. Financial backing has not yet been assured and the extent to which governments may be willing to provide additional support or guarantees has not been resolved. The cost is estimated to be about $12 billion.

There is, however, the other element of uncertainty which is less widely recognised. The incompleteness of the geotechnical research and the fact that methods of construction to counteract frost effects are as yet undecided, were referred to in the National Energy Board's report approving the Alcan proposal as well as in the Berger Report. In the case of the Mackenzie Valley proposals the significance of these deficiencies was overshadowed by social and economic considerations as well as possible environmental disturbance.[6] Because the proposed Alcan gas line follows the same route as the already constructed Alyeska oil line in Alaska, and a route in Canada through the southern Yukon, the questions of social and environmental disturbance are seen to be less important. The resolution of the geotechnical problems of a large-diameter gas line in the cold regions are, therefore, especially urgent. The successful completion of the Alyeska oil line may suggest that northern pipeline construction is well within the abilities of modern technology; but the fundamentally different problems that are associated with the proposed burial of a *gas* pipeline in the cold terrain of the North, must not be overlooked.[7]

The failure to recognise the profound nature of the outstanding geotechnical problems, is illustrated by a view prevalent in Canada; that because there is only some 150 km of discontinuous permafrost along the Alcan route in Canada, the research can safely be left to the Americans. Indeed this shorter *Canadian* 'permafrost' section is often cited as one of the advantages of the Alcan route, over the Mackenzie Valley route. The argument is parochial in the extreme (to describe it as 'nationalistic' seems less appropriate). The viability of the Alcan project as a whole depends equally on the successful construction of the Alaskan portion which traverses hundreds of miles of permafrost terrain, and the Canadian portion. The interest in a successful outcome, with minimum environmental disturbance, and the responsibility for ensuring this, rests with both countries.

As a consequence of the construction of the Alyeska oil line, much of the Alaskan portion of the route is now exceptionally well-described. With respect to design and construction, the usual requirements of preventing thawing and settlement of permafrost will apply for ancillary structures, such as roads, buildings, and compressor stations. The gas pipeline itself will, it is assumed, contain gas at temperatures colder than the ground in permafrost or discontinuous permafrost areas, leading to the problem of freezing and frost heave. There are some physical features of the Alcan route which are without counterpart in the Mackenzie Valley route. The rugged mountains of the Brooks Range in northern Alaska present a challenge, but this and other mountain crossings pose problems which, if different, are probably no greater than those associated with the extensive, poorly drained Mackenzie lowlands with their innumerable streams and rivers. On the whole, with the exception of the earthquake risk which applies only to the Alcan route, the more difficult terrain situations – swamps, river crossings, unstable slopes, etc. – occur along both routes, although with different frequency. The Alyeska pipeline has shown that such problems can be overcome in the case of an oil line. But the problems crucial for an oil pipeline are different from those that will be crucial for a gas pipeline.

6.2 SOME GENERAL PROBLEMS APPLYING TO GAS PIPELINES

The explosive nature of gas, and the implications of a rupture of a gas pipeline, are considerations which favour a buried pipe. On aesthetic, or environmental protection grounds too, the buried pipeline is favoured. Burial affords protection from certain forms of damage to the pipe, and also to some extent mitigates the effects of a possible explosive rupture. But burial also poses certain problems. Because of the low density of the gas there is a strong tendency for a pipe to float. This tendency is much less in the case of an oil line (the oil being heavier), although for both oil and gas lines, the tendency increases with the diameter of the pipe. It is the strength, and weight, of the overlying and laterally adjacent soil which prevents the pipe rising to the surface. Weak soils, those which are saturated, or are largely organic material may not hold the gas-filled pipe in place, so additional anchoring devices may be needed over large stretches. Saturated soils are frequent in permafrost regions, and in the less cold regions peat bogs are widespread. The low weight of the gas-filled pipe reduces the pressure on the ice that forms in the ground below the chilled pipe, and thus increases the risk of frost heaving, as discussed in Chapters 3 and 5.

The proposal that the Alcan gas line parallel the Alyeska oil pipeline (Fig. 6.1) for much of its length, is an example of the *corridor* concept. This is the belief that once a right of way has been established for one facility to pass through previously undisturbed terrain, subsequent construction and development should be restricted to the same corridor, thus avoiding new problems and an extension of disturbance to new areas. Knowledge of terrain within the corridor region becomes of course progressively more detailed. But this advantage is offset by the fact that additional operations may interfere with existing onces. Blasting and other construction procedures cannot be carried out in very close proximity to the oil line, a new service road may be necessary and the gas line may have to be placed so far from the oil line that few of the existing structures could be utilised. The relevance of earlier site investigations would then also be limited. The two lines would in any case have to be placed sufficiently far apart that any leakage or other failure of the one, would not interrupt the operation of the other. Finally, at some points it would probably be necessary that the gas line cross over the oil line, which raises the question of a structure which would maintain a secure distance between the two pipes.

6.3 CREEPING SOILS, RIVERS, AND GLACIER-DAMMED LAKES

The rejection of the Mackenzie Valley Pipeline proposals, following Justice Berger's proposal for a ten-year moratorium, should not lead to overlooking the second volume of his report.[8] The report recommends specific controls and requirements, which a government agency would enforce if the pipeline were built.[9] Most of the geotechnical elements and considerations relating to terrain are applicable to the case of the Alcan line. For example with respect to slope stability, it would be required that '. . . the

Company shall avoid where possible areas of questionable slope stability',[10] and that detailed studies be conducted to determine the cause, as well as the terrain conditions, under which various types of slides, flows, soil creep and other slope deformations occur in frozen or thawing soils. A detailed annual inspection of all slopes that may be unstable, using techniques acceptable to the agency, would also be required. The kinds of slope instability which are of particular concern have been outlined in Chapters 2 and 3. They involve, for the most part, phenomena unique to the terrain in cold regions, and their importance for the pipeline can be appropriately reviewed in this context.

The slow solifluction movements, a centimetre or so per year, of a fairly shallow surface layer (at most 2 m) of the ground, do not pose a direct threat to the relatively large pipeline structure. But these soil movements, together with the shallow 'skin' flows or slides which often occur in association with freezing or thawing, can ultimately produce dangerous conditions. Pressure could build up on the buried pipe or, more frequently, support could be removed. Drainage patterns might be modified, leading to erosion problems, and icings might also develop. More difficult to predict, and to rectify, are the larger, so-called deep-seated, landslides, which occur especially where there are steep river banks. On some occasions these slides may involve shearing of a layer of frozen ground, and in this respect they are unique to permafrost areas. Uncertainty about the ground conditions involved makes remedial, or avoidance, measures more difficult.

Of all the forms of slope instability, least is known about the creep of ice-rich frozen soils, and this movement may occur to many metres depth. The Berger Report concludes: 'The Company's research into the creep of ice-rich slopes should continue until the Agency feels that sufficient information exists to make a confident engineering prediction of those slopes with a potential for significant creep movements'.[11]

'Creep' refers to a slow deformation of a material, and occurs at stresses well below those required to produce a break, or rupture such as would give substantial immediate movement. In Chapter 2 it was pointed out that also solifluction often has these characteristics. But the creep of frozen soil could take place to depths very much greater than the movements normally considered as solifluction. Everyday examples of creep are the deformation and wrinkling of lead on old roofs, the relatively fast process whereby snow comes to hang over roof eaves, and the bending of wax candles during storage.

The flow of ice in glaciers is the most apposite example in the present context. How much ice has to be present in frozen soil to make it behave somewhat like a glacier? Glaciers flow even when the surface has only a slight slope. The steeper a slope, the greater are the stresses that may cause the movements, but the essence of a creep movement is that deformation occurs, although more slowly, even when the stresses are small. In the case of ice-rich frozen soils, there is firstly the possibility of the ice phase deforming, just as it does in a glacier, and flowing slowly while carrying the mineral soil along. Secondly, the fact that ice and water exist side by side, and that their proportions change when the temperature changes, or when the stresses on the soil change, means that small movements of moisture and soil particles can occur, which collectively may produce creep of the frozen soil mass. Indeed the same microscopic processes are involved as in the long-term migration of moisture which gives slow frost heave. The details of the ways in which such movements would be controlled by the angle of slope, temperature and stress gradients, and other factors, have not been established. But it is reasonable to assume that this kind of internal instability, arising because one component of the soil is at its melting point, would make a material liable to creep. It is unusual for an engineer to have to work with a construction material at its melting point, and in this respect the geotechnical engineer is obliged to do something no aeronautical engineer would consider for a moment.

Rock glaciers are the one feature already recognised as almost certainly the product of creep movements in frozen material. A rock glacier is a large accumulation of soil and boulders which, in its tongue-shaped outline and flow-like wrinkling, resembles a small valley glacier. When viewed from aircraft, they are reminiscent of thick, cold, poured porridge. Observations in France, Switzerland and Alaska have established that they sometimes move downhill at as much as one metre per year. They are fairly common in the Yukon, notably in the vicinity of the pipeline route. The interior of rock glaciers may be either a core of pure ice or soil cemented by ice, and the characteristic movements probably cease if the feature thaws.[12]

In general, the lack of field observations relating to creep phenomena is such that it is not yet possible to predict where and when such movements will occur. Consequently, to meet the requirement quoted above, monitoring systems would have to be set up on all those slopes where conditions would suggest the possibility of the movements. It is only rarely, as in the case of the rock glaciers, that there is a clearly recognisable surface form revealing the likelihood of movement. Thus, identification of those slopes which need to be monitored is difficult, and in the current state of knowledge the only generally useful characteristic appears to be the ice content of the ground. Even then, exactly how much ice constitutes a critical amount is not known.

It was not until some 20 years ago that it was generally realised that solifluction can occur on slopes which show no outward sign of such movements.[13] Further research may establish that creep of soil in frozen slopes, even to a substantial depth, is a rather general phenomenon. A build-up of frozen soil by creep movements which pressed on a pipeline would be serious. The foundations would be disturbed and displacement of the pipe might occur.

The Berger Report appeared before the Alyeska oil pipeline came into operation. Thus it is relevant to ask the extent to which the experience gained with that line should remove the concerns and uncertainties considered by Berger in his terms and conditions for a gas pipeline. Three points can be made. Firstly, if the gas line is buried and the gas chilled, the pipeline becomes locked in frozen ground. It might therefore be more susceptible to the effects of differential stresses or ground movements than the oil pipeline.[14] When buried, the oil line is surrounded by a thawed annulus, and when not buried, it is supported above ground with substantial provision for movement. Secondly, it is too early in the life of the Alyeska pipeline to say with certainty what problems are going to arise. Thirdly, rectification of the effects of soil displacements, around or involving the vertical support members of the Alkyeska line will be easier than when such displacements occur in the vicinity of a buried pipe. It is, of course, precisely where the most ice-rich and thus unstable soils occur, that the Alyeska line is elevated.

River crossings may involve burial of the pipeline below the river bed, or alternatively the pipe may be carried above ground on a bridge or other means of suspension. For the Alyeska oil line, burial below rivers was initially favoured, but was frequently given up because of difficulties associated with river bed scour and related problems – in particular, the changes which would be initiated by the placing of the pipe. The problems are compounded in the case of the gas line especially if this is chilled. The gas-filled pipe, being buoyant, must be held down in the river bed and this would require provision of large weighting devices or anchors. The development of a frozen annulus around the pipe might help anchor it but would also disturb the river bed. Normally permafrost is absent below major rivers. The presence of a body of frozen ground in the river bed could lead to accumulations of ice, which might prevent winter flow and lead to large icings. The approach to the river involves a downward bend of the pipe. This must be located sufficiently far from the river as not to be exposed by subsequent bank erosion and channel migration. Alternatively construction works will be necessary to maintain the river in its present confines. To carry a gas line over a river on a bridge raises problems not met with in the Alyeska line. It appears undesirable that a gas line be suspended from a bridge carrying a highway – a situation quite acceptable for an oil line (Pl. 11).

A chilled gas pipeline placed in initially unfrozen soil is likely to disturb sub-surface drainage patterns because of the large obstruction to water movement presented by the pipe within its developing annulus of frozen soil. The possibility of ground water rising above the ground surface on the upslope side of the obstruction, with the attendant possibility of icings, is additional to the risk of increased frost heave on account of the abundant water. Such ponded water also reduces the stability of slopes.

All these problems relate to the pipeline and its immediate surroundings. In certain places precautions against earthquake effects are required. There are some other questions which come to light only when the terrain is viewed in a wider context. One concerns the possibility of the pipeline being carried away in a flood, the origin of which is totally unconnected with the pipeline itself. In the southern Yukon the Alcan route approaches and borders the St Elias Mountains for some 200 km. There are large ice fields on the mountains with valley glaciers which extend outwards towards the lower land. These valley glaciers are commonly 10 or more kilometres from the highway and the proposed pipeline route, and do not therefore seem to be a hazard. But in the vicinity of the valley glaciers and extending many kilometres away, there are flat bench or terrace features, which are the remains of the shorelines of old lakes. The lakes were the result of damming up of rivers by the glaciers.

The Donjek Glacier, for example, has to advance only a short distance to block the Donjek river, and cause a large lake to be formed upstream. Sooner or later, perhaps in only a year, the glacier-ice constituting the dam would collapse allowing the lake to empty and the ensuing flood would probably reach the highway-pipeline route many kilometres downstream.

There are a number of such potential lakes, some large and some small, around the St Elias Mountains. A few of them, such as the Donjek river one, could cause serious problems although careful monitoring of the glaciers would provide adequate warning of the need for counter measures.

A different situation exists near Haines Junction. This place is surrounded by the shoreline of an old but recurring lake (Fig. 6.2). The Alsek river here flows southward (away from Haines Junction) and it is easy to see that the Lowell Glacier by blocking this river would cause the water to form a lake, which on the evidence of past shorelines would extend far beyond Haines Junction. There is a not inconsiderable risk of inundation in the next decade or two. Evidence from the age of trees, and the relative absence of vegetation remains[15] indicates that this lake last existed less than 175 years ago. Glaciers advance and retreat in association with even small changes of climate. Observations in several countries indicate that the period of glacial retreat during the first half of this century, when several kilometres of land was often exposed by the decrease of glacier size, is largely over. In recent years indications of a new advance have been detected in several glaciers. Sometimes there is a greatly accelerated movement, a surge[16] which is another phenomenon not fully understood. Even so, monitoring would give some warning.

Careful and continuing studies of the glaciers of the St Elias Mountains would avoid any risk of an unexpected disruption of the pipeline. But these ice-dammed lakes are an

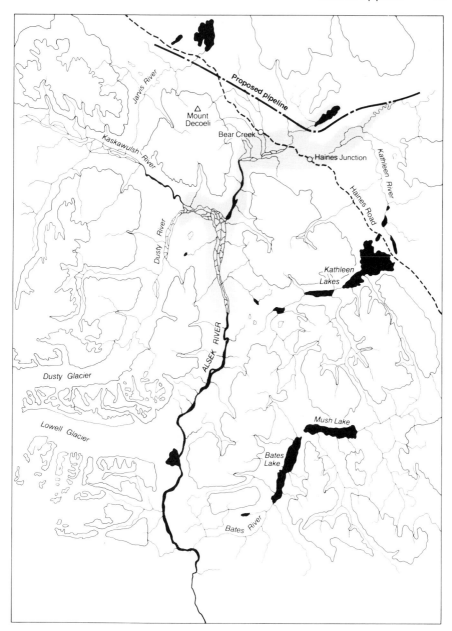

Fig. 6.2 Extent of lake about 150 years ago, formed by blockage of southward flowing Alsek River by Lowell Glacier. The approximate position of the shoreline is superimposed on a current topographic map. It is apparent that a small advance of the Lowell Glacier would cause the lake to reform, submerging the gas pipeline. (Part of Nat. Topogr. System, Canada sheet 115, St. Elias, reproduced by permission.)

example of the need for an intimate knowledge of the surroundings of the pipeline. The need for further research was emphasised by the interim report of the Environmental Assessment Panel[17] hastily established by the Government of Canada, when the Alcan route became a leading contender for the gas pipeline from Alaska. That report also stressed the role of ice in the soil and the need for studies to deal with the problems of freezing or thawing of the soil beneath the pipe.

6.4 THE ALCAN PIPELINE AND THE FROST HEAVE PROBLEM

The National Energy Board's tentative approval of the Alcan pipeline[18] appeared just when the complexity and significance of the frost heave problem was being increasingly realised.

Arctic Gas had had only a short time, just prior to the rejection of the Mackenzie Valley proposals, in which to suggest alternatives to the 'berm' concept for restraining frost heave. These alternative suggestions included the use of heating elements buried around the pipe to counteract the cooling effect of the cold pipe on the surrounding ground. The difficulty would be to maintain the appropriate amount of heating, such that neither freezing nor thawing of the surrounding ground took place to a significant degree. Overshadowing such proposals was the uncertainty as to exactly how often, and to what extent, frost heaving and frost heave pressures would be a problem.

It is now accepted that frost heave generates pressures at least equivalent to that represented by the weight of a 20 m deep soil mass. There have been further experimental measurements which confirm such pressures.[19] In addition, it is now realised that the presence of ice layers at great depths in permafrost, where the layers presumably formed with simultaneous heaving, is evidence of the magnitude of the lifting or heaving forces. If the ice layers (compare Pl. 1) are able to prise an opening for themselves when under the weight of tens of metres of overburden, it is clear that a similar process could occur in the soil a short distance below the cold pipe.

Unfortunately this kind of evidence provides little information about the *rate* at which the heave might occur when such high pressures are simultaneously involved. Rates of ice segregation, of ice lens formation, which may be very important in providing geological explanations of various terrain and ground forms, are not necessarily important in relation to the lifetime of a pipeline. Most experimental studies of frost heave, on the other hand, are carried out over hours or days and have been inspired by the problems of winter freezing of highways. To extrapolate the results of such studies over 30 years is obviously speculative.

Recent research of a more fundamental and theoretical nature[20] supports the concept of a slow, continuing water migration and ice layer formation *within* the frozen soil (that is, away from the boundary between frozen and unfrozen soil), with the development of correspondingly higher pressures (Fig. 3.4). Such research should ultimately allow calculation of the rates of moisture migration to, and within freezing soils, for various temperatures, soil and other conditions. Indeed several mathematical models have been devised[21] for use with computers, but their accuracy has not yet been confirmed.

An additional mechanism by which water may pass through an ice lens has been demonstrated by experimental studies in both the United Kingdom and the United States.[22] Ice is a polycrystalline material, and ice in frozen soils is normally composed of columnar crystals a fraction of a millimetre across, aligned perpendicular to the ice layer as a whole. Where several crystal faces intersect, there is a tiny column or channel of water present at temperatures near 0°C. The reasons relate to the Clausius-Clapeyron and Kelvin relations; at only slightly lower temperatures the cross-section of the water-filled channel decreases so that at a few tenths of a degree below 0°C there is very little water left. But at temperatures very near to thawing point, it appears the ice layers are so porous and permeable that water passes through them in significant amounts. It has been calculated that in one year, the equivalent of a layer of water 0.5 m thick passes down through a glacier by this means.[23] Even though it would generally be relevant only for a thin layer of soil where the temperature is near to 0°C, this porosity of the ice represents another mechanism to explain the penetration of water into the frozen ground in spite of apparent blockage by ice layers. One final report can be added to the accumulating evidence for frost heave of already frozen soil. Ross Mackay, using a simple extensiometer device, has observed what appears to be 1 to 2 cm of heave during one winter in already frozen soil.[24]

It may seem surprising in view of all the accumulated evidence, that there is still little agreement about what is actually going on in frozen soils, and in particular, disagreement about the precise, quantitative description of the process of freezing and thawing in porous materials. Doubtless there are those who will take exception to the picture of that process presented in this book. There is, however, more general agreement that the implications for the gas pipeline require serious study.

The fact that heaving continues in already frozen soil (if this is not too cold) is important with respect to how much of the length of the Alcan line might be susceptible to frost heave. While frost heave was considered in all the major reports and inquiries, the problem was usually seen in connection with the eventual freezing of *unfrozen* soil.[25] That is, the problem was thought to relate only to the discontinuous permafrost regions, and then specifically to unfrozen ground which the cold pipeline would pass through and cause to freeze.

However, because the high heaving pressures are generated in frozen soil (albeit rather 'warm' frozen soil), then the threat of heaving extends to the chilled pipeline when buried in *existing* permafrost. Initially the heaving problem was most feared at transitions between permafrost and unfrozen ground in the discontinuous zone. The pipe, so it was thought, might be 'locked' in an existing permafrost body, such that heave of the adjacent ground, freezing for the first time, would tend to shear or bend the pipe. In fact such transitions are unlikely to be particularly significant as soil freezes below the pipe; changes in the nature of the soil would seem to be far more important. Every soil scientist knows how soils change over even a few metres, from silts, to clays, or sands, and indeed through an infinite variety dependent mainly upon their geological origin. To predict, or even speculate upon the rate or amount of frost heave over any distance is therefore extremely difficult. In the case of the 'warm' Alyeska line, the main site investigation requirement was the location of permafrost bodies, and of ice masses in the ground; this was a massive task. But to predict the locations of future frost heave and the formation of ice masses is much more difficult.

The final designs for the foundations of the Alcan pipeline in ground prone to heave are not yet known. Indeed, there will probably prove to be no single solution. Various tentative proposals include placing the pipe in a wide ditch filled with coarse rock material (which does not experience frost heave); the use of a berm composed of such material and enclosing the pipe within it; and the use of insulation materials around the pipe, with or without heating devices to prevent cooling of the soil. The idea of an above-ground construction is still not entirely rejected. The length of the pipeline to be chilled may also be changed. Currently the southward cut-off point for chilling is only

65 km into the Yukon. A warm pipe of course is prone to problems arising from the thaw of surrounding permafrost.

The Report of the Federal Power Commission study of the various alternative routes for pipelines, did not concern itself with frost heave to the same extent as the concurrent Berger Inquiry. While pointing out the need for further investigations, the Commissioners were confident an adequate solution to the frost heave problem would be found.[26] This leaves open the question of how long it might take to find the solution. Later in the same report, the frost heave problem is seen as soluble 'with sufficient expenditure of design, time, and capital'.[27] With construction scheduled for the early 1980s, time is the item most obviously being spent. It is also apparent, as the Commission reports,[28] 'each applicant [to build the pipeline] will be operating at the margin of current technology'.

NOTES

1. Federal Power Commission (1977). Recommendation to the President, Alaska Natural Gas Transportation Systems.
2. Federal Power Commission (1977), p. 2. The Canadian government responded with, as some saw it, an untimely alacrity. Legislation was passed four months later in the House of Commons, approving the overland Alcan route through the southern Yukon.
3. There must for example have been a good reason why the 'United Association of Journeyman and Apprentices of the Plumbing and Pipefitting Industry of the United States and Canada' in their American section favoured El Paso and the tankers. Their Canadian counterpart favoured the Alaska Highway pipeline route (reported in National Energy Board, 1977, Vol. I, Chap. 1, p. 56).
4. National Energy Board (1977).
5. Northwest Alaska Gas Pipeline Co. is responsible for the Alaska section of the Alcan pipeline. In Canada, Alberta Gas Trunk Line (Canada) Ltd., Westcoast Transmission Co. Ltd., and Foothills Pipe Lines Ltd. make up a consortium, with Foothills Pipe Lines (Yukon) Ltd. being responsible for construction of the Yukon section (the 'northern' section).
6. Essentially Berger proposed a 10 year delay to allow settlement of native rights in the Mackenzie Valley. The further condition that no pipeline be allowed in the northern Yukon, however, effectively prevented the supplying of Prudhoe Bay gas to the Mackenzie line (Fig. 5.1).
7. Gas lines are in any case more demanding because of the explosive nature of the gas.
8. Berger (1977), Volume 2.
9. The Northern Pipeline Agency was soon established for this purpose.
10. Berger (1977), Volume 2, p. 136.
11. Berger (1977), Volume 2, p. 138.
12. White (1976).
13. Anders Rapp, whose detailed work (Rapp 1960) was mentioned in Chapter 2, measured movements in such slopes by observing the displacement of markers. In 1960 he demonstrated these experiments to a field symposium of the International Geographical Congress.
14. This matter is complicated. It could be that frozen soil at temperatures near 0°C, would be more capable of yielding slowly (by creep) thus relieving the differential stresses, than the unfrozen gravels which were packed around buried pipe sections of the Alyeska line. What is clear is that more investigations are needed.
15. R. G. McConnell, quoted in Kindle (1952).
16. Post (1969) – and other articles, especially in the Journal of Glaciology.
17. This Panel's short interim report (Environmental Assessment Panel, Alaska Highway Pipeline, 1977) contrasts with the extensive studies initiated by the Mackenzie Valley Pipeline Inquiry. While concluding 'that a pipeline can be constructed and operated in an environmentally acceptable manner . . .' this is made conditional upon proper planning, effective designs and adequate mitigation measures. The report repeatedly refers to lack of information. Somewhat more lengthy studies have been made by the Alaska Highway Pipeline Panel (1977) – an independent panel funded by Foothills Pipe Lines (Yukon) Ltd.
18. National Energy Board (1977).
19. Penner and Ueda (1977).

20. Miller (1978).
21. Several are described in American Geophys. Union (1975, 1976).
22. Oosterkamp (1975); Nye and Frank (1973).
23. Nye and Frank (1973).
24. Ross Mackay (1978).
25. In volume II of his Report (Berger, 1977, p. 134) Justice Berger writes: 'Although it is generally agreed that ice lensing and heave can occur in soil already frozen, that is, behind the frost front, as well as in soil as it freezes, experts disagree on whether or not this phenomenon is of engineering significance to the pipeline over the long term. The study of water migration in frozen soils is a new area of research, and the disagreement results in part from a lack of scientific knowledge.'
26. '. . . more experimenting remains to be done to determine the best means of combating frost heave. We have no doubt an adequate solution can be found . . .', so concluded the Federal Power Commission (1977, p. 1–35).
27. Federal Power Commission (1977), chap. VII, p. 11.
28. Federal Power Commission (1977), p. I-38.

CHAPTER 7
PIPELINES, SCIENCE AND SOCIETY

This book has followed the historical development of construction technology in the cold northern regions, a development which has led to the large diameter trans-Alaska oil pipeline, the largest and most significant achievement to date. That pipeline is, however, just the beginning of an enormous development in the North, if current plans for future pipelines come to fruition. The magnitude and effects of the projects will be so great as to impinge upon many aspects of North American society and indeed upon the world community. It is appropriate to consider the involvement, both regulatory and supportive, that society should have in entrepreneurial activities of such significance.

7.1 FUTURE PIPELINES

The United States and Canada are on the threshold of a momentous investment in the first large-diameter northern gas pipeline, the Alcan line, an undertaking of the same order of magnitude as the trans-Alaska pipeline, and presenting a greater technological challenge. Already an extension is tentatively planned, that would enable gas from the Mackenzie Delta to be fed into the Alcan line. This 1 100 km extension, known as the Dempster Spur, would probably connect at Whitehorse and follow approximately the route of the Klondike and Dempster highways (Fig. 7.1). The Dempster highway, which is incomplete, is intended to link Inuvik and Whitehorse. The Dempster highway would, of course, be an asset in the construction of the spur pipeline, serving as the access road for construction and maintenance, and thus reducing the overall cost of the pipeline. The difficulties experienced in the construction of the Dempster highway are instructive and perhaps ominous. The highway has been the subject of controversy, from the point of view of its role, cost, and the intrusion upon undisturbed terrain which it represents.[1] The criticisms are not without some basis. The highway's completion has been much delayed, the constructed portion being used only by adventurous tourists. Its design and construction did not involve the high degree of sophistication which the complex thermal problems and high construction costs might suggest. In some sections the road alignment had to be changed, after the clearance of vegetation initiated the thawing of the permafrost and settlement, which a few years later gave rise to a series of elongated ponds (compare Pl. 8). The realigned sections were constructed by placing fill on top of the natural surface. In some stretches, thaw settlement has resulted in drainage ditches in the bordering terrain lying at a lower level than the culverts through which the water is supposed to pass under the highway. This is because the excavated ditches themselves have the micro-climatic characteristics, associated with water surfaces, which initiate thawing of the ice-rich permafrost.

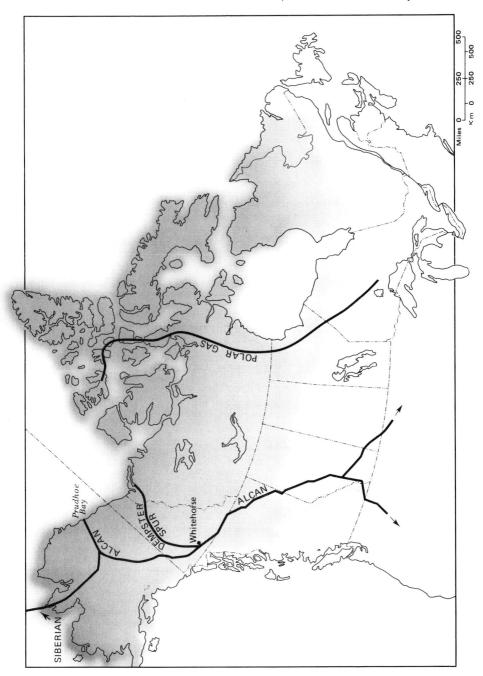

Fig. 7.1 Gas pipelines of the future?

The Dempster spur pipeline proposals mean that Canada would again be in the position (as it was when the Mackenzie Valley proposals were active) of planning and regulating a large gas pipeline traversing hundreds of kilometres of continuous and discontinuous permafrost. Irrational though it may be, the segment of the Alcan line in Canada which passes through discontinuous permafrost (some 200 km in the southern Yukon, near the Alaska border) is short enough that the weight of Canadian opinion would probably be in favour of leaving the basic research and development to American institutions, which institutions, it is tacitly assumed, would be self-sufficient. The attitude suggests, of course, an untimely shedding of responsibility because the Canadian section of the Alcan line would only function if the Alaskan section were successful. More important, such an attitude underestimates the need for, and the importance of, a wider understanding of the scientific and technological questions. The problems of the Dempster highway indicate the pitfalls arising from inexperience of even the simpler geotechnical questions; the problems are extensive, even though the solutions may not necessarily be difficult. The problems are reduced when there is a free exchange of information between groups in different countries, each group building on its own research and experience. The construction of the Dempster spur to carry only Canadian gas, would ensure greater Canadian participation in the geotechnical investigations and design.

Another massive project, which would also carry gas to the American market, is the Polar Gas line. Gas from sources in the Canadian Arctic Islands would be carried down the west side of Hudson's Bay, to meet the east-west trans-Canada pipeline (Fig. 7.1), which links with pipelines into the United States. Those parts of the Polar Gas pipeline on the mainland of Canada would be exposed to the same hazards as the Alcan line in its Alaskan permafrost stretches, or the Dempster line. The construction of the Polar Gas line is unlikely for at least 10 years, so that substantial experience would have been gained from the two earlier lines by that time. But the Polar Gas line would also involve a novel element – the submarine portions linking Boothia Peninsula, Bathurst Island and Melville Island.

The experience gained with submarine pipelines in the North Sea is not sufficient for high Arctic conditions. Little is known of submarine permafrost conditions, and significant studies only began to appear in 1971. In that year a paper in the American journal *Science*[2] described for the first time what were supposed to be submarine pingoes. The hills on the sea bed were detected by the echo-sounding device aboard the giant ice-breaker, the Manhattan. The origin of these features is still uncertain. They may be the result of 'eruptions' of water up through the sea bed, or they may be pingoes essentially similar to those of the Mackenzie Delta. The voyage of the Manhattan was a much-heralded attempt to show the feasibility of transport by tanker. The voyage did not lead to any positive conclusion. But if the existence of large numbers of uncharted sea bottom features, such as pingoes, is disquieting in relation to tanker traffic, it is even more so in respect to the laying of pipeline on the sea bed.

Grooves in the sea bed caused by the ploughing action of icebergs can occur even when the sea is hundreds of metres deep, although icebergs are unlikely in the vicinity of the Polar Gas line. More common are pressure ridges of winter sea ice which scrape the sea bed at up to 50 m depth,[3] as do floating ice islands which can reach even deeper. In the design and construction of the pipeline, the transition from subsea to the land would involve probably several kilometres of special engineering problems in the vicinity of each shoreline. The problems of shoreline change, drifting and erosion are similar to those experienced at river crossings. Yet the wave forces and the forces of floating ice are likely to be far greater than those experienced at any river crossing.

There is also a proposal for another submarine pipeline which would carry eastern Siberian gas under the Bering Strait to Alaska, where a further land pipeline would link with the Alcan line just north of Fairbanks (Fig. 7.1). In this way Russian gas would be carried through the Alcan line to augment supplies in California.[4] If that particular pipeline is at present an uncertain proposition, it is nevertheless clear that there will be expansions of the already substantial system of pipelines in Siberia. This also has some international interest. Although experience with permafrost is greater in Russia than anywhere else, the Soviet pipeline authorities purchase much pipeline equipment, especially for compressor stations, from Britain, France and Japan. Perhaps this reflects a longer-standing tradition of international development of pipeline technology compared to the more recent interest in problems specific to permafrost.

The extent of permafrost in China[5] is not well known, at least to those unfamiliar with the Chinese documents and publications, but there is an accelerating interest in pipeline technology in the cold regions of that country. The Chinese are also seeking both expertise and material equipment from other countries. Thus the techniques and problems of pipelines in cold regions are of world-wide concern. Indeed some 10 or 20 years from now the extraction and transport of gas and oil by pipeline in the Antarctic may become an international matter. There are thought to be enormous reserves there, possibly four to five times greater than those of the Alaska north shore. A number of countries have at various times made territorial claims in the Antarctic which have not received general recognition. In 1959 twelve countries signed the Antarctic Treaty which established wide rights of access but did not provide jurisdictional procedures should economic exploitation be proposed.[6] In 1977 the countries agreed to a moratorium on oil exploration and extraction while recognising that costs of extraction and transportation with current technology would be prohibitive.[7] The Spitzbergen islands and their vicinity, north of Norway, is another area where access by many nations is guaranteed by treaty, although the area is under the jurisdiction of Norway. As yet the lack of oil or gas discoveries has meant that the political questions have not been faced.

Clearly the rate of development of oil and gas pipeline networks in the various cold regions of the world will depend on many factors. One of these is the extent to which the technological difficulties arising from the cold environment can be overcome. This depends largely on the conditions for the necessary research and development being provided. International exchange in the form of sales of equipment, particularly those directly associated with the operation of the pipelines themselves already happens; but exchanges of information about the natural environment and the geotechnical problems that follow, are more limited.[8]

7.2 EARTH SCIENCES, EDUCATION, RESEARCH, AND THE STATE

The petroleum industry is dominated by companies which are multi-national. Names such as Shell, Phillips, British Petroleum, or Standard Oil, suggest massive organisations of great power and influence, backed by huge financial resources. It comes as a surprise to find that the success of projects of the magnitude of the northern pipelines is so dependent upon the sometimes fortuitous solving of scientific and technological problems in a limited period of time. Demonstrably, the construction of large pipelines in the cold regions is not achieved in the most efficient manner possible. The drastic change of the designs for the Alyeska pipeline due to unforeseen difficulties were a factor in the cost escalation.[9] More important, the successfully constructed line is now widely believed to be substantially 'overdimensioned'. The provision, for example, of costly thermal VSMs, often driven to over 15 m, is probably far in excess of that necessary; it was dictated by uncertainty about the nature and long-term behaviour of the ground. Remedial and maintenance costs for the entire undertaking remain to be shown.

From the point of view of the application of scientific principles relating to the process of freezing and thawing, there seems no reason why most of the present limited understanding of the behaviour of the ground in the North could not have been reached 50 years ago. It was not dependent upon the genius of an Einstein (as perhaps lasers, or space travel were) nor even upon the contribution of a Nobel Laureate. Some of the most recent studies of heat and moisture flow in freezing soils are based upon the work of Onsager in the late 1920s.[19] In the interval, freezing of soils was so little understood, that the relevance of his work went unnoticed. What was required was an impetus that would lead the engineer to the rather academic studies of geologists and geomorphologists who were exploring the natural terrain, or to the studies of physical chemists which were being applied in other fields, but not yet to the behaviour of soils in cold climates.[11]

Some areas of knowledge suffer from the inability of different kinds of scientists to get together. Medicine, for example, may need atomic physics, engineering may need biology. Those institutions, the universities mainly, where studies in many different fields are carried out in parallel are often subject to professional discrimination and academic sectarianism, and the much-vaunted inter-disciplinary approach then remains secondary to the preservation of established subdivisions of knowledge. In Canada, the country with the largest permafrost regions outside Russia, inquiry into the physical and thermodynamic properties of freezing soils has for the most part been made by a few scattered researchers. In the last decade there has been an expansion of studies of a biological, or social nature, relating to the North. But as far as the physical environment is concerned, studies have been largely descriptive, increasing the appreciation of the general nature of the North, but not providing the detailed analyses that are the basis of geotechnical competence.[12] In the absence of such competence, of course, a host of environmental and socio-economic problems may arise. In contemporary universities, and some industrial enterprises, research is mainly financed by the taxpayer's money, channelled through various agencies of government. Clearly, this has not resulted in the detailed studies necessary for application to northern pipeline projects.

The second volume of the Berger report on the Mackenzie Valley pipeline proposals,[13] puts forward measures to protect the natural environment which should be enforced by a regulatory agency, and similarly, measures to mitigate the adverse consequences of the pipeline construction for local residents. The measures to protect the natural environment would involve ongoing study of, for example, changes in animal populations. Substantial controls on construction activity are proposed to reduce disturbance to vegetation and natural terrain. In the section on geotechnical considerations, monitoring procedures are specified relating to possible frost heave of the pipe, settlement due to thaw, soil movements, whether slow creep, or rapid sliding, disturbance of natural drainage, and possible floating of the pipe. An important element in this monitoring is that it relates largely to quantities or effects which it is not yet possible to predict or explain. The monitoring procedures are thus not routine checks: they are instead a form of research. To this degree, the gas pipelines would be, like the Alyeska oil line, an experiment.

It would be unfair to criticise the petroleum companies for this risk and uncertainty except insofar as they are unwilling to admit the extent of the problem. The research efforts prompted by the needs of the pipeline designers are necessarily rushed and short-term, if expensive, operations. Previously most of our understanding of natural phenomena or of scientific fundamentals has happened quite slowly, as the result of studies that were made for their intrinsic interest, and undertaken only when the intellectual climate was propitious. Industrial enterprises carry out only that research which will provide the results necessary within the limited time available. Perhaps this is better described as development rather than research. The industrial scientist is constrained by the needs of the moment, and is not free to pursue the unexpected leads upon which real discovery depends.[14]

It is illogical to suppose that a company or companies dedicated to the building of a novel pipeline, will pursue with vigour a search for every possible difficulty that may conceivably (or perhaps, inconceivably) arise. Indeed that doubting attitude is the antithesis of the courage and daring associated with the great achievements of civil engineering through the centuries. So pervasive is the petroleum industry that it is not unreasonable to draw an analogy with the situation of the food industries. We do not expect the producers of foodstuffs to supply us with knowledge of all the physiological, and possibly harmful, effects which might follow from consumption of their products. Nor do we expect that governments merely lay down regulations restricting the use of foodstuffs on the basis of hearsay. We see it as the function of governments to sponsor and promote research to establish the harmful effects, if any, and their origin. Furthermore, it is the function of the state-financed education process to provide sufficient trained people for the purpose.

In the same way, the state ultimately has the responsibility of ensuring a pool of knowledge, and of scientific talent, for the entrepreneurial undertakings which may on the one hand be essential for survival of society in its present form, and on the other, may carry the seeds of immense destruction to our environment and well-being. In the case of the pipelines and northern resource extraction, the vicissitudes of the pipelines constructed, or envisaged, suggest that the forces of science and scholarship have not been properly marshalled.

7.3 CONCLUSION

In this book certain aspects of the nature and behaviour of the earth's surface in cold regions have been examined. These aspects have been shown to have much significance in northern geotechnical engineering generally, but especially in the design, construction, and continued operation of large pipelines. In tracing the history of the gradual application of scientific understanding to northern construction procedures, it has become apparent that both the development of such knowledge and its utilisation have been erratic and often slow.

The period of scientific application and of the development of rational construction methods in the north, has lasted only a few decades. During this time, however, there have been advances in basic science which have not found applications in northern engineering procedures until many years later. There have been delays because the findings of one breed of scientist have failed to reach another. There have been pitfalls and setbacks, but all the while the magnitude of the collective scientific and technological work which encompasses the cold regions has been small by comparison with efforts and achievements in other scientific fields. This is the more remarkable when it is remembered that those regions of cold and of permafrost where huge geotechnical projects are being undertaken lie for the most part within the boundaries of the world's most developed countries. The problems concern sudden demands for knowledge relative to huge areas of the northern lands – demands which cannot be satisfied quickly enough. The origin of the problems can be traced back to indifference in the past. So long as the northern and permafrost regions were to be explored rather than exploited, it could not be expected that 'applied' or technological studies would be promoted. Ironically, the lack of such studies may lead to the worst aspects of northern development.

Presently, however, there seems to be an additional explanation for the disproportionately small and scattered research effort relative to these areas. In modern society the direction of activity of scholars and scientific researchers is, in a collective sense, much susceptible to the influence of governments.[15] It is only governments, and in some respects large individual corporations or those associated with them which can, as in wartime, initiate or expand scientific activities in a manner effective in meeting sudden, massive technological demands. But these institutions in turn are responsive to public attitudes. Currently public attitudes are characterised by a questioning of the need for growth, and of the acceptability of disturbanace of natural environments, and a concern for the possible negative results of ever-larger engineering projects. Governments have, in recent years responded to these attitudes, by detailed examinations through commissions and enquiries. Such examinations have indeed been the impetus for many scientific studies, but the studies they initiate often bear the mark of haste, and are often of a descriptive rather than creative nature. The role of public enquiries and examinations is of course that of preventing or mitigating the effects of misdirected technology. There is an important obligation upon governments which is not widely recognised. They should, once objections of principle have been overcome, encourage the development of the knowledge and expertise necessary to ensure the viability and acceptability of giant projects such as the northern pipelines.

At the beginning of this book the question of what constituted success for major northern pipelines, was raised. It might be, as stated by the Federal Power Commission in its report[16] that 'there is virtually no chance that any system would become so costly as to be uneconomic – although some would see the risks as greater'. The quotation reflects the high prices expected for gas and thus the extent of the errors of design, the inefficiency, and the technical shortcomings, that could be absorbed before the project became 'uneconomic'.[17] In this respect the more 'economic' a proposal, the less is the constraint on ill-thought-out and potentially harmful design and construction. Consequently there must be additional criteria for the success of projects so wide-reaching in their implications.

The tacit assumption that modern technology can achieve something so apparently simple as a long tube carrying oil and gas may explain the indifference to the scientific challenge such a construction actually represents. The intellectual ferment instead centres on the socio-political, economic, or conservation issues. The experienced engineers and scientists tend to be employed by those striving to build the pipelines. The entrepreneurs, the multinational companies, are aware of the need for research. But exciting and well-managed though such research may be, vested interest in the construction of the pipelines will lead to neglect of certain questions. The companies tend to regard scientists in universities and research laboratories, both government financed and private, as antagonists or objectors. Thus the gap widens. The checks that society demands on engineering and entrepreneurial activities, in the interests of conservation and the general betterment or at least protection of the human environment, mean that knowledge must be accessible to society as a whole. Even if the findings and expertise of the technologists are made public, they will often be intelligible only to, and their significance fully understood by, those who themselves are involved with the activities of research and scholarship. It should be public policy, a responsibility of government, to ensure that there is a sufficient pool of knowledge to safeguard the public interest.

It is hoped this book has illustrated some of the excitement associated with investigating that frontier of research represented by the behaviour of the earth's surface and its materials in the cold regions. Such intellectual activity should be seen to be at least as important, in application to the questions of northern pipelines, as that of taking particular idealistic, perhaps dogmatic, standpoints for or against. With an increase of public interest in and support of the quest for knowledge, governments and industry might work together to ensure a sound and dependable technological and scientific base for northern development. Pipelines have become part of the North. Whether their future development will become ultimately a matter of regret and recrimination, or a favoured means of energy supply, must surely depend on the depth of our understanding of those pipelines' interaction with the particular northern environment.

NOTES

1. Berger (1977) makes a reference to the lack of assessment of the effects of the Dempster highway on the natural environment, in particular the effect on caribou migration patterns.
2. Shearer *et al*. (1971).
3. Personal communication, R. Wahlgren. Further references are Wadhams (1977), and Milne and Smiley (1978).
4. The input of Siberian gas would only be about 1 per cent of the total required annually in the U.S., by 1990 (Crowe, 1978).
5. The cold climate of parts of China is often overlooked. The highly continental situation and high altitude of the Himalayas and the Tibetan plateau is responsible for permafrost occurring even at the same latitude as North Africa or Texas (Ferrians, 1969).
6. The provisions for scientific research in the Antarctic are an unusual example of international cooperation (see account in Encylopedia Britannica, 1974). This followed from apparent economic unimportance of this region.
7. Hawkes (1977).
8. Apart from occasional personal exchanges between scientists of different countries, and some translation of scientific literature in Russian and English, the main exchange is represented by the International Permafrost Conference. First held in the U.S.A. in 1963, the second was in the U.S.S.R. in 1973, and the most recent in Edmonton, Canada in 1978. The latter attracted some 300 scientists from 13 countries. The proceedings of these conferences present the best overview of the state of the science.
9. In this connection it is worth asking to what extent government-imposed delays may have avoided construction based on faulty designs. Alyeska itself (TAPS, 1977) argues that the delays were the *cause* of much of the cost escalation.
10. Onsager received the Nobel Prize for Chemistry in 1967, although the work for which it was awarded was essentially completed in the early 1950s. Its application to freezing soils concerns the interdependence of flows of heat and moisture, and the development of equations to describe these flows.
11. There are many scientific papers on the effects of cold climates on soils. A review paper in 1944 (Troll, 1944) listed 1 500 titles. Probably less than 1 per cent drew upon fundamental physical laws in analysing the soil features.
12. In the period 1973–1978, only 12 per cent of articles appearing in the *Canadian Geotechnical Journal* related to the behaviour of freezing or thawing soils, or to other aspects of northern geotechnical problems.
13. Berger (1977), Vol. 2.
14. Much has been written about the role of serendipity. If there were no scientists left free to pursue their chance findings, or their own hunches and inclinations, there would be no fertile base for applied science.

15. The control may be of a very negative kind. Far from seeing university research as an essential element in national research and development, governments tend to see it as a political tool. For example, in response to ill-founded pressures, the U.S.A., Britain and Canada have all recently instituted discriminatory fees so that 'foreign' students have to cover a greater proportion of the costs of their studies, than those who study only in their home country. The net result for the three countries is to reduce the international transfer of knowledge and technology, and to penalize financially the more adventurous students (if they can still afford to study abroad). See note 8, Chapter 5.
16. Federal Power Commission (1977), Chap. 1, p. 38.
17. Already by 1978, the market situation for gas had changed so much that the cost of Alaskan gas in the U.S. would almost certainly have to be partly covered by the prices set for gas from other sources.

Alaska Highway Pipeline Panel (1977). *The Transmission of Prudhoe Bay Gas to American Markets: A Preliminary Environmental Comparison of the Canadian Arctic Gas Pipeline and the Foothills (Yukon) Pipeline in the Yukon and Northwest Territories*, 420 pp.

Allen, L. J. (1977) *The Trans-Alaska Pipeline*. Alyeska Pipeline Service Co., 2 vols.

Alyeska Pipeline Service Company (1977). *Summary Project Description of the Trans-Alaska Pipeline System*, 20 pp.

American Geophys. Union (1975). *Proc. Conf. Soil-Water Problems in Cold Regions*. (Spec. Task Force, Div. of Hydrology, Am.G.U.) 211 pp.

American Geophys. Union (1976). *Proc. 2nd Conf. Soil-Water Problems in Cold Regions* (Spec. Task Force, Div. of Hydrology, Am.G.U.) 185 pp.

Berger, Thomas R. (1977) 'Northern Frontier, Northern Homeland' *The Report of the Mackenzie Valley Pipeline Inquiry*, 2 vols., 213 pp. & 268 pp.

Beskow, Gunnar (1935) *Tjälbildningen och tjällyftningen med särskild hänsyn til vägar och järnvägar*. Stockholm, 242 pp. Statens väginstitut, Stockholm, Meddelande 48. (Also published as *Sveriges geologiska undersökning*. Avh. och Uppsats. Ser. C, 375, and translated into English: Soil freezing and frost heaving, with special application to roads and railroads: Tech. Inst. North Western Univ., Evanston, Ill. November 1947.)

Brown, R. J. E. (1964) *Permafrost investigations on the Mackenzie highway in Alberta and Mackenzie District*. Tech. Paper 175, Div. Build. Res., Nat. Res. Council, Canada.

Brown, R. J. E. (1967) *Permafrost in Canada*, Geol. Surv., Canada, Map 1246a.

Brown, R. J. E. (1968) *Permafrost investigations in Northern Ontario and Northeastern Manitoba*. Tech. Paper 291, Div. Build. Res., Nat. Res. Council, Canada.

Brown, R. J. E. (1970) *Permafrost in Canada*, Univ. of Toronto Press, 234 pp.

Burt, T. P. (1974) 'A study of hydraulic conductivity in frozen soils', M.A. thesis, Carleton University.

Burt, T. P., and **P. J. Williams** (1976) 'Hydraulic conductivity in frozen soils', *Earth Surface Processes*, **1**, pp. 349–60.

Considine, D. M. (ed.) (1976) *Van Nostrand's Scientific Encyclopedia* (5th edn), 2,370 pp.

Considine, D. M. (1977) *Energy Technology Handbook*, McGraw Hill.

Craig, R. F. (1976) *Soil Mechanics*, Van Nostrand Reinhold, London, 275 pp.

Crowe, Marshall (1978) Interview in *Maclean's Magazine*, 24 July.

Crudek, T. and **J. Demek** (1970) 'Thermokarst in Siberia and its influence on the development of lowland relief', *Quatern. Research*, **1**, pp. 103–20.

Embleton, M. and **C. A. M. King** (1976) Periglacial Geomorphology.

Encyclopedia Britannica (1974) 'Antarctica', *Macropaedia*, vol. 1.

Environmental Assessment Panel, Alaska Highway Pipeline (1977) *Interim Report of the Environmental Assessment Panel to the Honourable Romeo Leblanc, Minister of Fisheries and the Environment*, 55 pp.

Federal Power Commission (1977) *Recommendation to the President*. Alaska Natural Gas Transportation Systems. (Reprinted April 1978, by U.S. Dept. of Energy, Federal Energy Regulatory Commission.)

Ferrians, O. J., Jr., R. Kachadoorian and **G. W. Greene** (1969) *Permafrost and related engineering problems in Alaska*, U.S. Geol. Surv. Prof. Paper 678, 37 pp.

French, H. M. (1976) *The Periglacial Environment*, Longman, 309 pp.

Gamble, D. J. (1978) The Berger Inquiry: An impact assessment process, *Science*, **199**, pp. 946–52.

Gold, L. W., G. H. Johnston, W. A. Slusarchuk and **L. E. Goodrich** (1972) 'Thermal effects in permafrost'. *Proc. Can. Northern Pipeline Res. Conf., Tech. Memo. 104*, Nat. Res. Counc., Canada, pp. 25–45.

Harlan, R. L. (1973) 'Analysis of coupled heat-fluid transport in a partially frozen soil', *Wat. Res. Res.*, **9**, pp. 1314–23.

Harlan, R. L. (1974) 'Dynamics of water movement in permafrost: a review', *Proc. Workshop Seminar, Can. National Committee, Int. Hydrol. Decade*, pp. 69–77.

Hawkes, N. (1977) 'Science in Europe: Moratorium on Antarctic oil at October meeting', *Science*, **198**, 4318, pp. 709–12.

Hoekstra, P. (1966) 'Moisture movement in soils under temperature gradients with the cold-side temperatures below freezing', *Water Res. Res.*, **2** (2), pp. 241–50.

Institut Merzlotovedenia, im. V.A. Obrucheva (1953–57). 'Materially po laboratornym issledovaniiam merzlykh gruntov', (Izd.) *Akad. Nauk SSSR*, Moskva, Sb. 1,2,3.

International Permafrost Conference (1978) *Proc. Third Conf.* 2 vols. National Research Council of Canada.

Kindle, E. D. (1952) *Dezadeash Map-Area, Yukon Territory*, Geol. Surv. Canada, Memoir 268, 68 pp.

Lachenbruch, A. M. (1970) *Some estimates of the thermal effects of a heated pipeline in permafrost*, U.S. Geol. Surv. Circular 632, 23 pp.

Mackay, J. R. (1970) 'Disturbances to the tundra and forest tundra environment of the western Arctic', *Can. Jour. Earth Sciences*, **7**, pp. 420–32.

Mackay, J. R. (1973) 'The growth of pingos, Western Arctic Coast, Canada'. *Can. Jour. Earth Sci.*, **10** (6), pp. 979–1004.

Mackay, J. R. (1978) Personal communication.

Mackenzie Valley Pipeline Inquiry (1975) *Proceedings at Inquiry.*

Miller, R. D. (1970) 'Ice sandwich: functional semi-permeable membrane', *Science*, **169**, pp. 584–5.

Miller, R. D. (1978) *Frost heaving in non-colloidal soils*, Third Intern. Conf. on Permafrost, **1**, pp. 707–13.

Miller, R. D., J. P. G. Loch and **E. Bresler** (1975) 'Transport of water and heat in a frozen permeameter', *Soil Sci. Soc. Amer. Proc.*, **39**, pp. 1029–36.

Milne, A. R., and **B. D. Smiley** (1978) *Offshore Drilling in Lancaster Sound: possible environmental hazards*, Inst. of Ocean Sciences, Patricia Bay, Dept. of Fisheries and Envt., Sidney, B.C., 95 p.

Mitchell, B., and **R. Turkheim** (1977) 'Environmental impact assessment: principles, practice and Canadian experiences', in R. R. Krueger and B. Mitchell (eds.) *Managing Canada's Renewable Resources*, Methuen, 333 pp.

Muller, S. W. (1945) *Permafrost or Permanently Frozen Ground and Related Engineering Problems*, U.S. Geol. Surv. Spec. Rept., Strategic Eng. Study no. 62, 2nd edn, 231 pp.

National Energy Board (1977) *Reasons for Decision, Northern Pipelines*, 3 volumes.

Nye, J. F., and **F. C. Frank** (1973) 'Hydrology of the intergranular veins in a temperate glacier', in *Symposium of Cambridge – Hydrology of Glaciers*, Publ. 95, Intern. Assoc. Sci. Hydrol., pp. 157–61.

Oosterkamp, T. E. (1975) 'Structure and properties of ice lenses in frozen ground'. *Proc. Conference on Soil-Water Problems in Cold Regions*, Calgary, Canada. (Special task force, Division of Hydrology, American Geophysical Union.) pp. 89–111.

Péwe, T. L. (1975) *Quaternary Geology of Alaska*, U.S. Geol. Surv. Prof. Paper 835, 145 pp.

Penner, E. and **T. Ueda** (1977) 'The dependence of frost heaving on load application – preliminary results', *Proc. Intern. Symp. on Frost Action in Soils*, University of Lulea, Sweden, **1**, pp. 92–101.

Post, Austin (1969) 'Distribution of surging glaciers in western North America. *Jour. Glaciol.*, **8**, 53, pp. 229–240.

Poynting, J. H. (1881a) 'Change of state: solid-liquid', *Phil. Mag.*, Ser. 5, **12**, pp. 32–48.

Poynting, J. H. (1881b) 'Changes of state: solid-liquid', *Phil. Mag.*, Ser. 5, **12**, p. 232.

Rapp, A. (1960) 'Recent development of mountain slopes in Kärkevagge and surroundings, Northern Scandinavia', *Geogr. Annal.* XLII (2–3), pp. 71–200.

Roscow, James (1977) *800 Miles to Valdez*, Prentice-Hall, 277 pp.

Shearer, J. M., R. F. MacNab, B. R. Pelletier and **T. B. Smith** (1971) 'Submarine pingos in the Beaufort Sea', *Science*, **174**, p. 816–18.

Slipchenko, W. (1972) *Siberia 1971, A Report on the Visit of the Honourable Jean Chrétien, Minister of Indian Affairs and Northern Development and Official Delegation to the Soviet Union*, Information Canada, 124 pp.

Smiley, T. L. and **J. H. Zumberge** (1974) *Polar Deserts and Modern Man*, Univ. of Arizona Press, 173 pp.

Stearns, S. R. (1966) *Permafrost (Perennially Frozen Ground)*. Cold Regions Science and Engineering, Pt. 1, Sec. A2. U.S. Army Cold Regions Research and Engineering Lab., 77 pp.

TAPS (1977 ? – undated) *A Synopsis of Engineering and Cost Factors*. Prepared by owner companies of the Trans-Alaska Pipeline System, 31 pp.

Troll, C. (1944) 'Strukturböden, Solifluktion und Frostklimate der Erde'. *Geol. Rundschau*, **34**, 545–694.

Tystovich, N. A. (1975) *The Mechanics of Frozen Ground* (trans. from Russian), McGraw-Hill, 426 pp.

Wadhams, P. (1977) 'Characteristics of deep pressure ridges in the Arctic Ocean', *Proceedings, Fourth Intern. Conf. on Port and Ocean Engineering under Arctic Conditions*, Memorial Univ., Nfld., **1**, pp. 544–55.

Washburn, A. L. (1973) *Periglacial Processes and Environment*, Arnold, 320 pp.

Weast, R. C. (1978) CRC Handbook of Chemistry and Physics, 59th Edn. Chemical Rubber Company.

Weller, G. and **B. Holmgren** (1974) 'The microclimates of the Arctic tundra', *Journ. Appl. Meteor.*, **13**, pp. 854–62.

White, S. E. (1976) 'Rock glaciers and block fields, review and new data', *Quatern. Res.*, **6**, pp. 77–97.

Williams, P. J. (1957) 'Some investigations into solifluction in Norway', *Geogr. Jour.* CXXIII (1), pp. 42–58.

Williams, P. J. (1967) *Properties and Behaviour of Freezing Soils*, Publication 72, Norwegian Geotech. Inst., 120 pp.

Wynyard, John (1967) 'Roads and Pipe-lines', Wheaton, Exeter.

INDEX